DEDICATION

We dedicate this to our husbands. Their service in VW-3 DET A and VQ-1 (now the oldest squadron in the Navy carrying on its initial mission) in adventures far more perilous than ours, made it possible for us to experience our own most unique lifetime adventures. We especially must credit Stretch Varner for the initial impetus back in the early 1990's. In addition, we give our particular thanks to Pauly Varner for seeing that this project was completed through her patient thorough persistence.

— Wives of The Ferret Flyers

ACKNOWLEDGEMENT

August 22, 1956: while on a patrol mission from NAS Iwakuni, Japan, this VQ-1 P4M-1Q aircraft disappeared at night after reporting an attack by hostile aircraft 32 miles off the coast of China (near Wenchai) and 180 miles north of Formosa. There were no survivors of the 16-man crew. Photo USN.

Pete Bohley has been gracious enough to provide us with the title of our book. We are very pleased with his kind wisdom that is so appropriate to what our husbands and the crews on those Martin P4M-1Q *Mercators* were doing during the Cold War. The following story written by Pete explains his part in the crew and we wives wish to express our gratitude to him.

"On June 25, 1950, the North Korean Army invaded South Korea with the goal of reuniting the country under the control of the Communist North Korean leader Kim Ill Sung. Within weeks, the South Korean Army was in full retreat. Its American "advisors," scrambling to help, could do little to stop the impending loss of the entire peninsula. As the battle lines contracted around the southeastern port city of Pusan, help began to arrive. Struggling to recapture some of the capability that had won WWII, complete collapse was narrowly avoided, and by October, the North Koreans had been pushed back above the 38TH Parallel. Some of our nation's top leaders feared that further incursions into North Korea would cause the Chinese to enter the "war" and ignite World War III. Others chose to push for complete destruction of the North Korean Army and control of the entire country.

It is not surprising then, that with this concern the knowledge (or lack thereof) of the defensive nature of this part of the world came under intense scrutiny. One of the Navy's responses to this need was to deploy updated surveillance aircraft. The Squadron eventually became known as VQ-1 (Electronic Countermeasures Squadron One). The Navy chose the Martin *Mercator* P4M-1 as the basic platform

East Asia

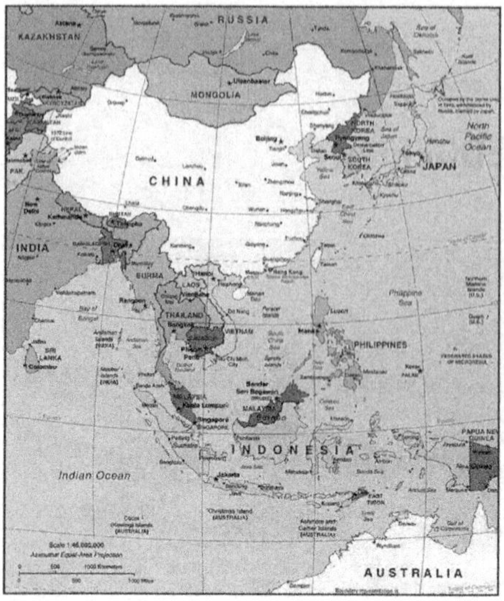

From the Republic of the Philippines, the crews of VQ-1 covered Asia like Sherman-Williams Paint "covered the world."

and extensively modified it for the new mission. It was newly outfitted with state-of-the-art electronic monitoring equipment (some left over from WWII) and designated the P4M-1Q. The aircraft was a unique four-engine long-range plane. It had two reciprocating engines and two jet engines. It was well suited to its "ferret" (ferret – to search and discover) job of electronic surveillance. The squadron was tasked with searching out and recording every electromagnetic signal that could be found from Hainan Island (near Viet Nam) to Vladivostok (on the Russian coast north of Japan). The method was basically simple. Simple and dangerous.

The men flew at night — on very dark and moonless nights. The aircraft was unmarked and emitted no radiation, neither visible nor electromagnetic. The three gun turrets were manned and the guns were loaded. The crew of the aircraft would settle in for a ten-hour session along the China coast in search of "information." It was not always easy to find.

The newer radar facilities could be shut down when these "unknown" aircraft were spotted by shore-based search radars. Fire control radar was often not operating at all unless a threatening approach was used. A threatening approach sometimes produced interceptors who searched for the intruder and sometimes found what they were looking for. Several P4Ms were shot up and two were shot down, one with the loss of all hands, during their decade-long service in the Far East.

As a young, single, Naval Aviation Electronics Technician, I joined the Squadron in the fall of 1953 and began flying as a radioman on Ferret missions. It was a very exciting time for me and many like me. As we prepared for a mission, we loaded up with a flight suit; Mae West (life vest); parachute harness; parachute; .38 caliber pistol; Escape and Evasion Kit which included survival equipment; gold piece, blood chit (a silk banner printed with the words "Bring this guy home and we'll pay you" – in several languages), IF4 canned food packages (that thrill wore off quickly); and climbed into the nose turret with those two big 20mm cannons practically in my lap, headed on toward adventure! I still look back on those times fifty-five years ago as some of the most memorable of my life.

I had no concern about those who might be worrying about me (my parents), as they didn't know what I was doing, or that it might be dangerous. At that time I

P4M *Mercators* on the maintenance ramp at NS Sangley Point, R.P., circa 1954.

had no awareness of those in the crew who were married, and whose wives (who only knew enough to be afraid for their husbands) not only lived on the Base (NAS Sangley Point, Philippine Islands), but were close enough to the runway so they could hear the aircraft taking off in the dark for parts unknown. When they woke up early in the morning, they hoped to hear that familiar sound of a P4M bringing their men home, but they never really knew when, since we could be gone for as much as a week or longer.

I am grateful to those wives who are sharing their stories with us in these pages. It was a part of the story that I didn't realize existed."

Pete Bohley

PRC MIG-15 Fighter. Photo CNN.

who lived the adventure with them — all who collectively gave birth to VQ-1 - are now but a memory. Not one P4M remains anywhere to testify to this important and colorful chapter in Naval Aviation. The night skies they flew are empty. High-flying specially-designed aircraft and spacecraft employing computer technology now accomplish a greatly expanded electronic mission. But VQ-1, the Squadron born of a handful of airmen flying the darkened P4M-1Q *Mercator* through night skies over foreign shores during the silent "Secret War," lives on as a viable and crucial asset for an expanded reconnaissance and countermeasures mission in defense of our great nation.

Dean North

DaLoris Reynolds

THE PHILIPPINES

"The Philippines! Way over there! When do we go?" I asked my husband Jack when he told me of his new set of orders in May 1953. Flashbacks of the Second World War, MacArthur, Corregidor, islands, coconuts, natives, and jungles all raced through my mind instantly. This sounded very exciting, yet I felt a grain of anxiety. Then I was informed our five-year-old son Jon and I would not be traveling as a family, but we would go later when housing was found for us. Meanwhile we would wait for about six months while we lived with my parents on their farm in Iowa.

In November of that year, my mother accompanied us to San Francisco where we would embark on the USNS *Barrett*. Very concerned about our welfare, she was relieved when we met up with Mildred Dean with her two children. She had been to Guam as a dependent and knew all the ropes about shipboard. Mildred and I bonded instantly.

Officer Landing, NAS Sangley Point, RP.

My bunkmate was Shirley Colton, a pregnant younger wife. Our trip consisted of prolonged twelve-day nausea for each of us and hospitalization for Shirley. The dining room was torture for me. Just the smell of food erased my

From a vintage post card. The back reads:
"Named in honor of Major Gen. Charles D. Barrett, USMC. The Barrett is the newest and most modern troop ship in the M.S.T.S. fleet. 534 feet long, 73 feet wide, and can do 20-1/2 knots. Used for carrying troops and dependents."

mealtimes and back to my bunk I'd return. Fortunately the ship's crew had much entertainment for the children and Jon was happy. I must admit at this point in my journey I thought, "No man in the world is worth this much agony." What a way to lose weight.

When we arrived in Manila, Philippines, my husband met us at shipside and we went to Cavite City, our home to be. So happy to see him, I forgot to be angry about the miseries of the *Barrett.*

Our new home was one of seven houses for American families in a compound known as Certeza Court. The compound overlooked an offshoot of Manila Bay known as Bacoor Bay and was surrounded by a cement wall eight feet high. On top of the wall were imbedded broken Coca Cola and beer bottles as well as strands of barbed wire to deter would-be burglars.

Besides the American families in the compound there was a house occupied by Loplop the caretaker and his wife Crazy Mary, a genuine casualty of the war that occupied these islands. Poor dear soul! The Japanese soldiers had taken her baby, thrown it up in the air, and caught it with a bayonet, killing her baby right before her eyes. It was too much pain for this mother and she lost her mind over the atrocity.

Crazy Mary was a petite five feet tall thin lady, very wrinkled, with no teeth. Most days she slept away, the nights were the most disturbing to her fragile mind. Dressed always in a white uniform and go-aheads, she paraded and patrolled the entire two-acre compound on foot all night. She chanted and did singsong talking in low tones that we could not understand.

A full moon was most disturbing to her. We would awaken to her voice right under our open bedroom window. She always squatted down against the house and stayed there, carrying on her conversation with an imaginary person. Her glistening black hair and white uniform in the full moonlight were an unforgettable sight. We became used to her way of life very quickly and would fall asleep again with her still talking under the window.

Crazy Mary was named by the Filipinos. We never knew her by any other name nor did we know any more about her background. Our compound was very safe because she terrified the local natives. We did learn how horrible war can be to the individual citizen mentally, and understood better the brutality of the Japanese occupation of the Philippines.

Jack had hired three Filipinos: one to cook, one to clean, and a houseboy. Unemployment was rampant and wages cheap in Cavite. At that time it was a status symbol for American wives to do no housekeeping. All employees were friendly and eager to please. However, they were not used to most of the household items of Americans and had to be trained how to use and maintain them. When our silver tea service was unpacked and needed polishing I found Placita using Brillo pads to remove tarnish from the coffee pot. How was she to know?

Groceries were bought at the Commissary on the Base except for rice, bananas, coconuts, papayas and mangoes. A trip to the local Sari Sari was one where every item was bargained for and quantities of rice were measures in gantas versus pounds. Missing from the commissary were fresh produce and fresh milk. Dried powdered milk mixed with water was our only milk. It tasted like chalk and the BX's (Base Exchange) attempted to make ice cream from it but that was a huge disappointment.

Due to the rainfall surplus most houses, called nipas, were built on stilts to keep dry. Even so, a persistent smell of mold from decayed vegetation mixed with food odors permeated the air throughout Cavite. One of my desires while living in Cavite City was to see the inside of one of these nipa huts, the traditional pole mounted home of most of the islanders. My wish came true in a way I had neither foreseen not imagined.

As I drove to the Base one day, a ten-year old Filipino boy playing ball in the street ran in front of my car. The Navy had warned us wives and husbands that should anything like this ever occur, we must with haste get the injured party to the Base medical unit. I complied with the orders.

An adult Filipino (I can't remember the relationship) lifted the lad into the car and a U.S. Navy doctor saw him as soon as we arrived on Base, about a mile and a half away. The boy had a broken leg. I was so relieved that a broken leg was the extent of the boy's injury.

About three weeks later I received a telephone call that the parents wanted to see me at their home about the boy's accident. Jack told me to decline and ignore the call. Well, my curiosity got the best of me, and I decided I would see the parents and not tell my husband until after the meeting. In my head I had already guessed it would be a shakedown for money. I was curious, however, and thought I would surely be okay just to drive up in front of the nipa hut and call to them from my car.

Typical Nipa Hut on the island of Luzon, Republic of the Philippines.

Without telling Jack, I arranged the meeting. When I arrived, a man in the street told me the parents were

Sangley boat landing

expecting me inside the nipa to talk. Should I? Should I not? Well, this was my chance to see inside the nipa, perhaps my only chance. Surely as a U.S. citizen I would be okay.

I vividly remember climbing the dark stairs up to the nipa and wondered why I had let myself in for this discussion! I was motioned to sit down in the main living room which seemed devoid of any furniture other than a few chairs. Almost immediately a group of men seemed to appear out of nowhere and surrounded me at a respectful distance, Instantly I knew I was in a 'Kangaroo Court' — here and now — in that nipa. It was time for me to keep cool and level-headed. Yes, they — all men — asked for $100. I don't recall seeing one woman. Trying to take charge of the situation, I stood up, told them I was not paying any money, and walked toward the door. No one laid a hand on me or blocked me as I descended the stairs toward the car. I was surprised they had not asked for more money but not so surprised as I was grateful for my safety. Yes, I had seen inside a nipa, but I cannot tell you to this day what it looked like except for a blur of chairs in an otherwise empty room devoid of color. The experience has stayed vivid in my mind through the years, as I count my blessings.

After a year in Cavite, our name came up to move on Base. Our new abode was a furnished Quonset hut consisting of a master bedroom, a second bedroom, living room, dining room, and kitchen right on Manila Bay. We felt privileged to be in such a nice Quonset, able to see Manila's night lights, and live next to the pier for the ferry boats.

We had a new employee, again named Placida but spelled differently. She was more experienced and a cook, and would be our only one now. We grew very fond of her.

She was very capable with Jon who was entering first grade. All the house girls on Base slept in one large Quonset hut. They wore floppies, or 'go-aheads' as footwear and left them at the Quonset door where they lived. Our Jon decided to be naughty early one evening. He went over to Placida's Quonset and threw all the floppies into the bay. Punishment deserved!

Jack and his squadron mates were either 'gone' or 'home' as far as the wives' discussions. We all understood the absolute imperative not to discuss their absences. When they arrived home we could hear the planes descending and landing from our nearby quarters.

One particular evening Jack had just taken off for a mission. I received a phone call from the plane captain's wife that a plane had just crashed at the end of the runway and it probably was Jack's and her husband's plane. Having lived the life of a pilot's wife and been associated with many crashes, I knew the SOP. The chaplain would be there pronto.

The *Margaret* is a 65-foot pleasure yacht, on loan to NAS Sangley for recreation and sightseeing trips to Corregidor, fishing trips, swim fun at small island beaches near the entrance to Manila Bay. No swimming in Manila Bay as it is very dirty, muddy and polluted.

Jon had been playing outside and was really dirty. The first thing I did was wash him clean. Just as I finished with Jon a phone call informed me it was not Jack's plane. Thank God!

The wives occupied time with Bridge, baby showers, trips to Manila via the Navy ferry, luncheon with speakers, and keeping the home fires burning. I went to Manila monthly to a beauty salon, finding Filipino very artistic and good hair stylists. This was a real treat for me!

Most couples took their R&R to Hong Kong for a shopping bonanza. Upon their return, friends would gather for a 'show and tell' to see all the inexpensive purchases and beautiful Oriental items. Hong Kong even had real ice cream in one store, certified to be safe by U.S. standards. For my husband and myself, our Hong Kong trip was the initial step to a lifelong love affair with this tantalizing metropolis.

Baguio is north of Manila in the mountains of Luzon. It was another R&R for most couples, especially for golfers. It was cool and delightful and an interesting trip up by auto, passing several checkpoints guarded by police to check identification. Security was tight due to the HUKS still present and active in certain areas.

A trip to the Dole Pineapple Plantation was always fun and only a one day's outing.

Ready-made dresses were not available. I wrote Frederick and Nelson in Seattle and they sent out a box of dresses for our selection. The ladies were happy to see the new fashions and something fresh in their closets. I have had a warm spot in my heart for Frederic and Nelson ever since.

Health problems? A couple! While in Hong Kong I contracted hepatitis and was bedridden and contagious for several months. Jack had an emergency of a blocked colon. He was sent to the Medical Hospital at Clark Air Force Base for surgery. The Navy flew me there in a helicopter to visit him.

Foreign languages have always been easy for me. I enjoyed learning Tagalog phrases that I still use today with our local Filipino medical personnel. Knowing that my husband's flights helped our nation's security kept me contented. I felt fortunate to learn about the Far East firsthand, plus making warm, lifelong friendships as a bonus.

DaLoris Reynolds

Jack Reynolds

DaLoris (Nihart) Reynolds

Autobiography

February 28, 2011

She died March 15, 2011.

I was born and reared as an Iowa farm girl.

After graduating from Iowa Wesleyan College, I studied for and secured a Radio Engineers license in Chicago.

I met my "Fly Boy" Naval Officer at a USO dance while I was employed as a commercial writer for the radio station in Ottumwa, Iowa.

We moved 37 times across the United States plus the Philippine tour.

Hobbies include geneology, antique art glass, and golf.

We just celebrated our 65th Anniversary.

Ruby Abbott

FROM THE "GARDEN SPOT OF THE MIDWEST" TO THE "PEARL OF THE ORIENT"

We moved every two years while my husband, Lt. M. J. (Bud) Abbott USN, was on active duty. He retired from the Navy after 23 years of service. I loved being a Navy wife as we traveled around the United States, going from one duty station to another. We had many opportunities to visit National and State Parks, and other interesting places.

Arriving at a new duty station, we would hunt for a house to rent. Then we cleaned it before moving in and then gave it another thorough cleaning when we departed for the next place to call home. Our children became accustomed to going to new schools as we moved from place to place. Sometimes they would meet up with friends they had not seen in several years. It was a great life.

Bath time in Cavite — young mother with toddler in washing pan

We met many interesting people and made many long lasting friends, especially the people while stationed at Naval Air Station (NAS) Sangley Point, Philippine Islands. Our husbands flew in a highly secret squadron VW-3 Det.A that later became VQ-1. We were not aware of where they went or what they did on those patrols, but we surmised

LT Bud Abbott

that it was dangerous and necessary for the safety of our Nation. Declassification of that information was made public a few years ago.

Prior to going to the Philippines Bud was a flight instructor at NAS Pensacola, FL, from Aug. 1950 to 1952. He then attended a six months Navy Photo School at NAS Pensacola. Upon completion of that school, he received orders to report to NAS North Island, CA for further assignment.

While in Pensacola our oldest son, Bruce (age six) attended first grade. Brian was four. I was pregnant and not looking forward to the long drive from Florida to California. Bud and I decided that he would go ahead to NAS North Island to find out where the "further orders" would be. Those orders were to NAS Sangley Point, Philippine Islands. I could not travel with him; housing was limited, making it necessary for me to wait for permission to travel, whenever that would be!

We sold our three-bedroom house in Pensacola and Bud drove us to Belvidere, IL. our hometown. The sign at the entrance to Belvidere announces that it is the "Garden Spot of The Midwest". It is about 80 miles Northwest of Chicago, IL

We were fortunate Bud's parents and my mother still lived in Belvidere, IL. They welcomed us warmly, telling us we could stay as long as we needed. The boys were excited about staying with my mother plus having the other grandparents nearby.

Bud left the car with me and flew to San Diego. He spent a couple of weeks at the Base for some training, and purchased a Ford to ship to the Philippines. After completing the necessary training, he drove the car to Oakland, CA for shipment and then reported to the Federal Building in San Francisco to get a Military Flight to the Philippines. That was the last time I talked with him until arriving in Manila, P.I. Letter writing was our way of communicating and letters were very slow in arriving.

I was a bit stunned about going to the Philippines but remembered learning a little in school about that far away place. I wanted to know more about where it was that I was to travel and then live for two years. I went to the library and found some old books discussing head hunters in the mountains. That did little to inspire me about traveling to the Philippines Islands.

Many decisions loomed ahead of me. I welcomed Bud's letters and wished that he could be with me to help answer all his questions. One letter arrived with the information that he had started the paperwork in motion for the children and me to join him. His first question was, "Did I want to go by ship or travel by plane?" I remembered that a plane load of dependents had recently crashed. I sent my letter back telling him that I would prefer to make the trip by ship.

Another letter explained that he was flying many patrols, so he was away from the Philippines a lot. The Base had no hospital, only a Dispensary. With that information, I decided to remain in Belvidere, IL until our daughter, Judith, was born. That would also allow Bruce to complete First Grade in June.

Letters from Bud arrived often and it was obvious that he missed us as much as we missed him. The boys have always been close to their father, and soon there would be another baby to join our warm and loving family.

The boys enjoyed the grandparents and even learned to ride two wheel bicycles that summer. Bruce made many friends from school and shared them with his brother Brian. However, the letters from dad were always special.

September 1st arrived and so did baby Judy. I did not feel like traveling right away so we had time to get all our physicals, shots, passports, and paperwork completed. Brian, age 5, was terrified of the shots and he tried to run away. I found myself chasing him down the corridor near the doctor's office to bring him back for his shots. The Navy advised me to send my Passport (yes, we had one passport with one picture of Bruce, Brian, Judy, and me) to the Federal Office Building in San Francisco. They would advise me when travel arrangements were completed. I sent a letter off to Bud telling him that we hoped to be there soon. We were to sail on the USNS Barrett February 9, 1954.

The doctor's office informed me that the Shot Card Records would be available for pick up on Fri. Feb. 3. Instead of four cards there was one card listing by date the shots given. I was shocked, and told the nurse that the Navy required one card for each person. She obliged and made out the individual cards.

On Sunday, Feb. 5, Bud's parents drove us to the airport in Rockford, IL to board the commuter flight to O'Hare Airport in Chicago, IL. It was a sad farewell for all of us, but the grandparents knew that it was best for us to be together as a family.

I had brainwashed the boys, now ages five and seven, that we would be traveling first by plane and then on a ship half-way around the world to find their daddy. I impressed upon them the fact that they must do everything I ask of them.

Off we went with our suitcases, my overloaded pocketbook, diaper bag, and books for the boys. The commuter plane ride was bumpy with elevator effects, and I was hoping for a better flight on United Airlines to San Francisco. At O'Hare, we walked and walked before I found the United Terminal. I told the boys to sit where I could see them while I went to the desk to check in. Brian refused to leave my side. The desk clerk was very nice and when she noticed the problem, she told me that I could board the plane right away. She helped gather up our belongings, corralled the boys and took us aboard the big plane. We had our choice of seats so I put Judy in a seat beside me and Bruce and Brian chose the two seats across the aisle. Soon the stewardess arrived with dry ice to keep the formula cold.

It was a beautiful clear sunny day and a smooth ride. The stewardess took Bruce and Brian to visit the pilots in the cockpit. The pilots gave them United

Airline Wings to put on their jackets. Before landing at San Francisco the pilot made a detour over Yosemite National Park and tipped the wings so we could see half-dome. It was beautiful.

A bus took us to our hotel near Union Square in San Francisco. The Navy offered to make the reservations for us if we sent them the money. It was an old hotel, but very convenient to shopping and sightseeing.

Monday I asked the hotel to call for a taxi. We had to check in at the Federal Office Building. I asked the taxi driver to take us to the Federal Office Building.

He said, "Where's that?"

I responded, "I don't know. That is all the address I have."

He thought a minute and then announced that he thought he knew where it was!

I was apprehensive!

Looking out at the sights as we drove thru the streets of San Francisco, it looked like we were in the Orient. I asked the cab driver where we were.

He said, "We in Chinatown. Big celebration here last week. Big parade, lots of fun!"

He was telling me all about the Chinese celebrations. I was trying to take in all the sights. In addition, I kept wondering where the Federal Office Building was located.

Rounding the next corner, we saw the humungous building setting way back from the street. There were many steps leading up to the massive doors. I asked the cab driver to wait for us until I checked to see if it was the right place. He told me that was not necessary.

He said, "Just come to curb and wave at a passing cab. They stop for you."

We climbed the stairs and entered the building. In the distance, we saw a desk where a lady sat. As we approached she said, "Good morning Mrs. Abbott, so glad to see that you have arrived." It startled me to be addressed with such familiarity. She handed me my Passport and papers with instructions on how to proceed thru the steps required prior to boarding the ship. The doctor took our temperatures, checked our smallpox vaccination, looked over the immunization shot records, and declared us ready to leave for the Philippines.

Before leaving the Federal Office Building, I asked the receptionist how she knew who I was. She said, "That was easy, I had your Passport with a picture, plus most everyone else has already checked in."

The next step was a trip to the Philippine Embassy to get a Visa to enter their country.

I found the address, entered the building and took the long flight of stairs to the second floor. I rang the buzzer and a beautiful oriental lady who was elegantly dressed came to the door. I explained what I needed and she took my Passport. A chair was located so I was able to sit down with Judy on my lap. Judy was five months old and a healthy baby, but getting heavy for me to carry. After what seemed a long time the beautiful lady returned my Passport with the necessary stamps on it.

Making our way down the long steep stairs was worse than climbing up them. Carrying Judy caused me to be a little off balance so I was very cautious. We finally reached the bottom and out the door into the beautiful California sunshine.

A short taxi ride to Macy's was our next destination. I bought a "Kiddy Koop" (a screened in crib) for Judy and a stroller. I had already sent items from Belvidere to the USNS Barrett for shipment to Manila.

On Tuesday, I wanted to see more of San Francisco, but taxis were getting expensive. I asked at the hotel desk about tours and they told me about Grey Line Tours. We could walk a block to Union Square and then cross it diagonally to the bus stop. Flocks of pigeons were being fed by elderly men seated on benches along the walkways. Bruce and Brian delighted in running around chasing the birds in hopes of catching them. The boys squealed with delight as the birds screeched while taking flight. They would have stayed there all day running after the birds, but I wanted to see more of the city.

The bus was a great way to see 'The City by the Bay'. We saw the Aquarium, Lombardi Street, Mission Dolores, The Cliff House, Height-Asbury St. (no hippies), and much more.

An interesting event happened on the tour. One of the ladies on the bus seemed to be interested in the children, and me. She asked me what I was doing in San Francisco and I told her that we were on our way to the Philippines where my husband was stationed. Her husband resembled Dr. Davis who had delivered my boys. Their features and red hair were very much alike. I asked her where they were from and their name. She told me they were from Freeport, IL but I don't remember her name. Freeport and Belvidere are neighboring towns. I then asked her if they were related to Dr. Davis. She said that she had been asked that question many times, but she did not think so. I later found out that Dr. Davis was from Freeport. It is amazing whom one meets on a bus tour.

The day finally arrived that we were to board our ship. After breakfast, we packed our suitcases and I called room service for help. We waited and waited for them to come, and I was getting concerned about being late getting to the Federal Building. It seemed that several people at the hotel needed help. Finally, they came and we got a taxi. This time the driver knew where the Federal Office Building was located.

Arriving at the Building two ladies came to the taxi and introduced themselves. Pauly Varner and Mary Jane Edixon told me that their husbands were also pilots in VW-3 Det A. Mary Jane was holding baby Joanne, and she said that they were calling my name. Pauly offered to take Judy and the boys and I went into the Building to see why they were calling me. It was a letter from Bud.

A loudspeaker announced that we could all go to lunch and be back by 1:00 p.m. to board the bus that would take us to the ship. If we missed the bus then take a taxi to the dock. All seven of us piled into a taxi and went to a place that Pauly knew about for lunch.

While at lunch, I opened the letter from Bud and found out that he was in the hospital dispensary at NAS Sangley Point with infectious mononucleosis. He didn't know when or if he would recover. I was stunned! The letter continues, "If you are still in Belvidere when you get this letter, maybe it would be best if you stayed." My mind is racing and I am very concerned about Bud. I wanted to be with him. What if the Navy had Med-evacted him back to the States and he wasn't in Manila when we arrived?

I looked at Pauly and Mary Jane and sensed that they already knew about Bud. Pauly told me that she had received a letter from her husband, Stretch, telling her to find me and tell me about Bud. She in turn had told the news to Mary Jane. I had never heard of the disease, but they had heard of it. They said, "He will be alright." Therefore, it was decided that I continue with the trip. They both agreed that the military would not allow me to continue to the Philippines if my husband was not there. This news was very upsetting for me, but with their assurance, I decided to board the ship.

We finished lunch in plenty of time to make it back to the bus for the 1:00 p.m. departure to the dock. Bruce and Brian spotted the USNS *Barrett* and were very excited about going aboard. I remember people selling newspapers with the headlines, USNS *Barrett* sails for the Orient!

When they escorted us to our cabins Mary Jane and I were thrilled to find that we shared a cabin. Two small cribs in between the bunk beds were provided for the baby girls and the boys climbed happily to the top bunks. It was a large cabin on the main deck.

Shortly after settling into our cabin, a loudspeaker announced that a meeting was to be held, and everyone is to attend. I left the children in the cabin promising them I would return soon. At the meeting, we learned that the ship was not going to sail. There had been a fire in the stacks and that needed repairs before the ship could go anywhere. The USNS *Barrett* would remain in dock until another ship was brought alongside. It would act as a floating hotel for all of us. Four days later the USNS *General M.M. Patrick* arrived and we walked from one ship to the other. It was a rusty old ship left over from WWII. This time Mary Jane and I did not share a cabin. I had a Navy Nurse with me who was headed for duty in Hawaii. When she left, we had the cabin to ourselves for the remainder of the voyage.

The USNS *M.M. Patrick* hauled up anchor and untied from the dock before sunset on Valentine's Day 1954. The motion of the ship gave me a queasy feeling, which I thought, was due to not having eaten all day. We did not go out on deck to watch from our ship departing San Francisco, and sail under the Golden Gate Bridge. I sure did not like being seasick. I wished I had never started on this long journey, and it was only a week

USNS *M. M. Patrick* in 1952.

ago that I left Belvidere.

Soon we were out in the open Pacific Ocean and the ship was not only going up and down with the forward motion, but also rolling from side to side. Everyone was sea-sick. Brown paper bags were placed on handrails along the passageways for people to use, but still there was vomit all over the decks. It was rough all the way to Hawaii. It took the ship seven days from San Francisco to Hawaii. The nurses advised us not to drink liquid, but to nibble on crackers to help settle our stomachs.

They also told us to go out on deck. One day we did make it to the top deck and it was awesome watching the swells and see the ship rolling from side to side. After four or five days, we began to feel better.

Marge Arthur, a friend, and her 2 daughters met us in Hawaii. She wanted to take us on a tour of the Island, but I still wasn't feeling very well so I did not see much of the area. Marge said I probably had land sickness.

During our first week, at sea we had our first Fire drill. This entailed getting the boys into large size life jackets, fastening and snapping all the necessary gear. Then I had to get into my large size, old dusty and dirty life jacket before putting baby Judy into her vest. The weather was hot and humid which made us feel very icky, but out on deck we went. Everyone had an assigned space near a lifeboat. We listened to the instructions and then returned to our cabins where we took off the jackets and washed up. It was a most uncomfortable experience, but it is required for everyone on the ship. Everyday enroute to Hawaii the Fire Drills continued. I hoped we would never have to abandon ship in a real emergency.

Departure time from Hawaii was 4:00 p.m. and the sea remained rough for a few more days. Our ship could only make about 10 knots so we were in for a very long voyage to the Philippines.

Sailors worked on the ship everyday, chipping rust and painting. I hoped it would not spring a leak and sink. The USNS *M.M.Patrick* was a very old ship compared to the USNS *Barrett*. However, after hearing stories about the USNS *Barrett* we were glad that we did not sail on her. That ship was rejected by the American President Lines because she had an unusual roll. We did not need any-more motion than we had on the USNS M.M. *Patrick*. Luck was with us after all.

The children and I have made two trips by ship to the Orient and two return trips. I am not interested in any Cruise Ship experiences. I have had enough of life at sea. It is boring and sickening! Another thing I do not care for are Convalescent

Care Centers. I spent two weeks in a Rehab Care Center after my hip replacement surgery, and I hope I never have to be in one of those places ever again.

Our next stop was Guam where more of the passengers disembarked. Then it was finally off to Manila. At sea between Guam and the Philippines, a large airplane flew over our ship and made a circle over us. I wondered if it was Bud flying the plane. Later I learned that it was Bud and Stretch Varner but they did not think we were on that ship. It looked like an old rusty freighter to them, and they did not think we would be on a ship that looked so bad.

About a day before reaching Manila we noticed an unusual odor in the air. The ship was sailing thru the islands to get to Manila on the western coast of the island of Luzon.

Someone came up with an explanation of the odor. They called it "The Orient"!

It was a happy day that we docked in Manila. People on the dock were waving and shouting as they spotted loved ones on the ship. Then the people on the ship started shouting and waving back while waiting for the gangplank to be rigged for people to run across into the arms of their loved ones. I could not see Bud anywhere. Leaning over the side of the ship, I spotted Lt. Jerry Terry who I was not expecting to see in the Philippines, but we had known while living in Pawtuxet River, Maryland. While getting ready to debark Bud came aboard and found us. An amusing incident occurred when Brian whispered into my ear asking, "Is he really my daddy?" He was only four when Bud left us in Belvidere, IL a year ago. Bud's appearance had changed slightly. That was probably due to his recent illness. Bruce had no problem remembering his dad! I then introduced Bud to his new daughter Judith, a beautiful blue-eyed blond six-month-old baby.

We bid a happy farewell to the USNS M.M. *Patrick* as we piled into the Ford Coupe for the drive around Manila Bay to Cavite City and our new home. While driving Bud said, "Wait till you see our new home! It is called a Nipa Hut." He pointed out Bamboo huts on stilts telling us they were Nipa Huts. We also saw caraboa lying in water ditches alongside the road. There were wagons and carts drawn by caraboa or an occasional horse. That was my only trip by car around Manila Bay. After that, we always crossed the bay by boat. That boat from NAS Sangley Point also stopped at the U.S Aircraft Carriers, which anchored in Manila Bay. That Bay is so shallow the Carriers cannot go all the way to the docks. Therefore, the personnel on board use the boat transportation to get to and from shore.

Boys use a caraboa for transportation near Cavite City.

Our house was not a Nipa Hut! It was one of two houses surrounded by a stonewall with a gate large enough to drive thru. The yard was sand except for two concrete runners for parking the car. No one lived in the other house. A Sari-Sari store was across the dirt road in front of our house, but I never went there. I do not know what they sold in that store.

Next door, on the other side of the stonewall lived a family that had a pig tied near their back door. A big kumquat tree shaded their yard. Another large tree was used by young Filipino boys to climb so they could see into our yard. I think they were curious about our American families. I did not like living there. I wanted to be near other American families so I would have someone with whom I could talk.

All military and dependents in the Far East changed 'Green Money' for MPC (Military Payment Certificate). It looked somewhat like Monopoly money. The 1-cent, 5-cent, 10-cent, and 20-cent were also paper, but a little smaller than the dollar bills. We could exchange MPC's for Filipino Pesos and Centavos. This was necessary to prevent black market dealings. We used MPC for shopping on Base, and Filipino money in town and for paying the hired help. There was an office at the Base Exchange where we could cash checks and exchange money. I do not recall any bank on Base.

In 1954, Americans could do no wrong. WWII had been over for nine years. The Filipino people vividly remember being liberated from Japanese soldiers who were very cruel. Our maid would often tell us how nice the American's were and how badly they were treated by the Japanese.

We stayed in that house for a little over two months when Bud and Stretch were assigned to take one of the planes back to San Diego, CA, and then to bring another plane back to the Philippines. They expected to be gone about four to six weeks.

I told Bud, "I am not staying in this house. I want to move to a Compound where other Americans live."

New Base housing was recently completed making it possible for several families living off Base to move into these new houses. This made more housing available for people living in Cavite City. A house in the compound where Jack and Ursula Gillooly lived was available. Jack was also a pilot in VW-3 Det. A. We arranged to rent that house and Bud found a couple of sailors willing to help and a truck. This made everyone happy, especially the boys who could now play with the Gillooly boys. We had green grass in this compound and from my upstairs bedroom window, I could see a large field where the local fishermen spread out their nets to dry and mend. I felt much better being here while Bud would be gone.

Pauly remembers coming over for dinner and play cards one evening. We had so much fun we neglected to watch the time. It was late and dark and she did not want to drive thru Cavite to her compound. We welcomed Pauly to stay the night with us. The gate on her compound was locked at 10:00 p.m. and "Crazy Mary" patrolled the area with a 45 pistol. If Mary thought you did not belong in the compound she would not unlock the gate, and Pauly would have been left alone outside the gate.

Bud and Stretch returned from San Diego with gifts they delighted in buying for us. Pauly and I were so happy with dresses they had picked out for us. The local seamstresses made beautiful clothes for us, but having dresses purchased from those excellent shops where the men could review the latest fashions paraded past them by live models.

Soon they were back to flying their highly secret missions, leaving for those one, three, five, or seven days at a time before returning around dawn. Life was never routine, but we kept busy with all our various activities. We partied late at night or early in the morning when the men drank San Miguel beer to unwind, and the wives drank coffee while cooking breakfast. Then it was off to bed for some much needed rest. It was a rather upside down life.

Our name for Base housing came up in June, and we moved into a two-story duplex with two bedrooms and a bath upstairs. The spacious downstairs had a wonderful kitchen, dining room, and living room. The carport was covered and had a large storage room that could be locked. At the back of the quarters was a lovely screened porch.

The big living room floor was parquet wood squares, and these were highly polished by the maid. She used a lot of Johnson's paste wax and polished it with a half coconut husk. Placing the husk under one foot she did her hip hop dance moving the husk back and forth across the parquet tiles. We did not remove our shoes when entering the house, but the maid always left her bakyas (native wooden shoes) outside.

Some families hired two people to help with the housework. In addition to the maid a family would hire a lavendera to wash and iron clothes, or Amah to take care of small children, or a cook to prepare meals, or a houseboy to take care of the yard, polish shoes, and keep the car washed and polished. We hired an Amah to take care of Judy and she was wonderful. Her name was Floracita Dugenia, but we called her Flory. We kept in touch with her for several years, but have now lost track of her. It was a nice way of life for a short period of time. The house was always sparkling, dishes washed, children happy, and I was somewhat happy, except for those mysterious flying missions that took Bud away from us.

We enjoyed the native fruit that our house girl, Fely, bought at the Cavite City Market. There were delicious mangos, huge papayas, many kinds of bananas, and coconuts. The Filipino rice was exceptionally good, and smelled like popcorn when cooking. The house girls ate a different kind of rice than they cooked for us. The closest type of rice we can find here in the States is Jasmine rice. The potatoes were not good so we did not buy them. Meat and chicken we purchased at the Base Commissary. Most of our vegetables came in cans as transport of fresh vegetables by ship took several weeks. The fruit from the native market was washed in a solution of clorox and water before eating. This was necessary to kill any parasites

We lived near the Headquarters Building at NAS Sangley Point. At 8:00 a.m. every morning, a Flag Raising ceremony was held. A bugler sounded the call as the

American Flag was raised. This was followed by the Station Band playing to the colors and other patriotic songs. At the conclusion of the ceremony the American National Anthem and the Philippine National Anthem was played. The amahs brought the pre-school children to watch this ceremony. This also gave these Filipino girls a chance to socialize with their friends.

One evening Bruce and Brian were playing in our backyard when we saw a flash of light. A neighbor also saw the flash and came to investigate. Brian had a small toy rifle with a metal barrel. Pushing it into the ground it hit an underground electrical box left by the workmen who built the houses. It was still attached to live electrical wires. Bud called someone who came and immediately and removed the hazard. Brian was unharmed as the grass was dry.

Mosquitoes were prevalent. Every evening a small tractor pulled a kerosene-fogging machine thru the housing area. Big clouds of the awful smelling fog emitted from the machine. The foggy mosquito repellant drifted thru the screened houses and surrounding neighborhood killing the bugs. It must not have been healthy for us to breath.

Our bathroom had a wood paneled wall and we noticed small holes in it and sawdust on the floor. Could it be termites in this new house? Yes, it was. A call to the housing office brought the exterminator crew to get rid of the termites. A short while later we noticed the same thing in the wood used in the glass front of the built in China cabinet. The termite exterminators returned and treated all the wood in the entire house. I wonder what is left of that house fifty-five years later.

Our telephone could be plugged into a jack upstairs in our bedroom. It was my practice to take the phone up there when I went to bed, but one evening when Bud was away on patrol I forgot to take it with me. During the night it rang and I rushed downstairs to answer it. Our windows did not have any glass windows, just screens with wood jalousies that we closed only when rainy or very windy. If I let the phone ring it would wake up the whole neighborhood. When I answered no one responded and that made me very nervous. I yelled at them, hung up, and took the phone upstairs. Later I told Bud about it, but he told me not to worry about it.

Later a friend who lived in a one story house told me she received a call late one night, and when she answered a male voice said, "I want you!" She quickly turned off the lights, checked the locks on the doors and took the phone to her bedroom. Later she could hear footsteps outside, but whoever it was left the area. This all happened while her husband was off on a flight. Living on the Base was not as secure as we had thought it would be.

Ants invaded our house. They came in at the backside of the living room, going along the baseboard and up the stairs to a window above the carport where they left the house. They looked like a moving black line. I showed them to Bill Deane, another VW-3 Det A Squadron member, who lived in the apartment on the other side of our building and asked him if he had seen any in his duplex. He sent to the States for some Grants Ant Stakes. When they arrived, he placed them all

These are the trucks used to take us on trips around Corregidor to see the destruction from WWII.

around our building. We never had another ant invasion.

One day I was stopped by the Marine Guard at the main gate as I was driving into Cavite. He advised me to close the car windows even though it was hot and humid.

I asked him, "Why?"

He said, "Look up the street. They are throwing buckets of water at people and cars."

I rolled the window up. Then I decided to speed by them. My gas pedal stuck and I could not release it. Jamming my foot on the brake I saw people scattering in all directions. Then the gas pedal released; but I was still shaking. I found out it was John the Baptist Day and everyone was being Baptized! They left me alone and I continued on my way.

Ursula Gillooly and I rode our bicycles on the hot dusty roads to see the Flagellantes. It was Good Friday and the Filipinos used Bamboo strips as whips to beat themselves on their backs until they were bloody. This was their way to show penance for their sins. It was too horrible for us to see. I took one picture and we hastened back to the Base.

At one of our reunions, several years later, we recalled our shopping trips to Manila with Irma Kruzich, another squadron wife. She was a wonderful Manila shopper and knew where to go to find bargains and interesting items. She would show us where to buy beautiful fabrics. She sewed beautiful clothes for herself in the latest of fashion. We always enjoyed our trips to Manila when Irma was with us. Someone recalled being near a church when she stopped us to let us know that there were beggars around the corner. She told us not to look at them or give them money or anything. They will fight over it and harass you until you give them more. We peeked around the corner. It was shocking! I never went to Manila by myself.

The Squadron scheduled a trip on the Margaret to Corregidor. The Margaret was a 60 foot yacht used by the Base for trips in Manila Bay. Arriving at Corregidor we climbed aboard old Army trucks to tour the island. We saw big cannons used by the American and Filipino Armies. Many were pointing skyward, as if shooting at planes. The trucks took us high on the hills where we could look down on what was left of the Army Post. The mile long barracks and parade grounds were in shambles. Trees grew in the bombed out barracks and the parade grounds were becoming overgrown by vegetation. We also stopped at the caves that were carved out of the hills on the island. It was spooky to walk into those caves and know that the soldiers left behind when Gen. MacArthur escaped, had been captured and tortured.

We took a wonderful trip to Baguio, the summer Capitol of the Philippines. Bud was able to get a four door sedan from the Base Motor Pool for our drive to the cool mountains of Baguio. He had reserved a cottage for us at Camp John Hay an R. & R. (Rest and Recreation) facility for service personnel stationed in the area. I remember looking back down the switch back mountain road we had just driven up from the lowlands below. It was awesome. We saw young girls and women washing clothes in the rivers. They would run and hide as we drove by them. We took Fely our maid with us.

At the Baguio market we bought matching horse heads carved from mahogany and asked them to drill holes in the heads so we could make them into lamps. We used them for many years. We also went to a French Boys School where they made silver filigree jewelry. We purchased a couple of necklaces.

Fely was not always able to converse with the people in Baguio. She spoke Tagalog, English, and a dialect called Ilocano. There are over 90 dialects in the Philippines.

This is one of the BIG guns on Corregidor.

Bud brought home two Japanese bicycles for the boys. They were equipped with hand brakes. The boys soon mastered the new brake system and enjoyed riding them around the Base. Bruce liked that bicycle the best of any that he has had. Bud and I had English bicycles with hand brakes instead of coaster brakes. The tires were small and inflatable. They were easy to ride bicycles. Bicycles were bought and sold among the families arriving and departing. When our husbands were away on their patrols they stay at other Bases to eat and get some sleep. Bud was able to bring the bicycles home on the plane.

Judy developed a heat rash on her neck when first arriving in the Philippines and it became infected. The doctor gave me erythromycin ointment to rub onto her skin. It soon healed. She also developed thrush in her mouth and refused to eat. I noticed the white film in her mouth when she yawned. The corpsman at the dispensary painted her mouth with gentian violet and I continued the treatment until it healed.

Many military wives played bridge and I was interested in learning the game. Kay Lewis, a Navy wife offered eight of us wives lessons in bridge. She cautioned us not to get involved with the advanced players who held tournaments at the O' Club. I have always been grateful to Kay for teaching me how to play bridge, and her timely words of advice. I still enjoy playing bridge.

Philippine history and what went on during WWII has been of great interest to me. I read the book "The Drainpipe Diary" by Theresa Cates. It has been reprinted and the new title is, "Infamous Santo Tomas." Theresa was of Hungarian parents and was an American Nurse. Her fiancé was a Pharmaceutical Representative in Manila when the Japanese dropped bombs on the city. They were captured and held as prisoners of war at the University of Santo Tomas until the American forces regained the Philippines and rescued them. They and all the other prisoners nearly died of starvation. I found Theresa and Catesy living in San Marcos, CA and we visited them briefly. They were true survivors. They are both deceased.

We left NAS Sangley Point in April 1955 to return to the U.S.A. Leaving NAS Sangley, we took the daily boat to Manila. Instead of going to the U.S. Embassy landing they took us to the Port Area. That was much closer to the ship we were to board for our trip back to the States. The small boat was manned by two Filipino sailors. They tied the small boat to a large wooden floating barge. Everyone had to climb out of the boat up onto the barge. Both were in motion because of rough waves. I was terrified and thought I would have to remain in the Philippines for the rest of my life. Bud took Judy and the boys and yelled for me to "Come on!" Finally in desperation I made the climb to the floating barge. For a few minutes I thought I was safe, but then realized I had to walk across the wobbling barge to a plank that I must walk across to the stable wooden dock. Bud with Judy and the boys walked "The Plank" successfully. I looked down and saw the dirty roiling water below the narrow plank that had nothing for me to hold onto while I crossed to the solid wooden dock. Finally, with Bud's demanding voice saying, 'Just walk across'. I made the attempt and was successful in reaching solid footing. Never did I ever imagine such a rude departure or ever dreamed of such scary conditions to endure.

My memories have taken me from Belvidere, IL, "Garden Spot of the Midwest" to Manila, "Pearl of the Orient," 55 years ago.

Ruby and Bud Abbot, 2011

Ruby (DeMunn) Abbott

Autobiography

February 26, 2011

I was born in Belvidere, Illinois and resided with my parents in the house where I was born. I attended Belvidere schools, graduating from high school in 1942 and was a cashier in the office of the local telephone company.

I married my high school classmate, Ensign Myron 'Bud' Abbott USN, a U.S. Navy pilot trained to fly multi-engine aircraft. We have two sons, Bruce and Brian, and one daughter, Judy (Abbott) Carney. My time was devoted to caring for my family. We moved every two years, living in seven States and the Philippines and Japan. For the past 47 years, we have lived in Lompoc, CA.

My children are senior citizens. My six grandchildren are young adults. They are Andrew, Kelli, Michael, Brooke, and Mark Abbott, and Jonathan Carney.

Eleanor (Lee) Chapola

MY LIFE IN THE FAR EAST
Before and During VW-3 Det A and VQ-1

I think that my story is rather unique even among the stories of the other women who lived in the Philippines during the early to mid 1950's when our husbands were stationed with the VQ-1 squadron. I lived in the Philippines not only in the 1950's but also in 1938-1939 when my father, Robert E. Bower, a lieutenant in the US Navy, received orders to be stationed there. I was sixteen years old and really did not want to go and leave my friends. It meant that my schooling would also be interrupted. I wanted to stay in Washington, D.C., where we lived, and finish high school at Eastern High School. My mother, Nora Larner Bower, would not

USS *Pecos* (AO-6) in 1938. Photo USN

Bubbling Well Road, Shanghai before WW II. The imposing building is that of the China United Assurance Society. Photo courtesy Virtual Shanghai.

USS *Augusta* (CA-31). As flagship of the Asiatic Fleet, *Augusta* was a key instrument of American naval power in China.

let me stay behind so we embarked on quite a journey.

We flew to San Francisco and took a President Line ship to, Philippine Islands, with stops at Honolulu and Yokahama, Japan. My father met us in Manila and we had quarters in a not-so-nice place in Manila. I think it was on street called A Mabini. I remember the bath being outside, a few steps from our apartment. It was quite different from what we had in Washington, D.C.

Mom and I really did not get out too much and I think she was nervous going about the city by herself. My father was stationed aboard the USS *Pecos* and was gone a lot of the time. We were in Manila for several months and then traveled first to Shanghai, China and then on to Tsingtao, China. In Shanghai, we lived on Bubbling Well Road. In both cities my parents had their own room and I had mine. Dinner was served communal style.

I met my future husband, Peter Chapola in Shanghai. He and a friend of his, Dick Enders, were on a tour bus at the same time that my mother and I were touring the Chapei area, which the Japanese had invaded in 1937. We said hello and talked a little. I did not see him again until we moved to Tsingtao. At the time Pete was stationed on board the USS *Augusta*.

One day, Mom and I were at a café on the beach in Tsingtao and Pete and a few friends walked in.

We talked, or I should say, he talked to my mother. He asked me to dance but I turned him down, so he asked my mother and she gladly obliged him. I was more interested in one of Pete's friends. Mom asked Pete to come join us for dinner at a later date, which he did, and then he started asking me to go for bicycle rides or to go to the beach. I think maybe he thought I was too young — I was 16 and he was 8 years older. Maybe he thought my dad would not approve. My mother really liked him.

It was an interesting time in China. There were people from all over the world who gravitated to China for one reason or another. There was a large population of Russians, many Germans, French and other nationalities. We had our own rickshaw driver, which was a great way to get around. One time when my father was due to arrive on the ship, our driver told me I had to go with another driver because, "Master is coming." I didn't want to go with someone else but he was insistent.

In Shanghai we lived four blocks away from the Fourth Marine barracks and the Scottish barracks were across the street from our boarding house. I thought some of those Marines were very cute. When we moved to Tsingtao the Japanese barracks were next door. When the Japanese soldiers marched on the sidewalk everyone would make sure they were out of the way. I didn't move for them and made them march around me or go out into the street. My mother was sure I was going to get killed but they were young and so was I and I did not think anything would happen to me. Some of the soldiers would approach me when I went to the beach and ask to take my picture. I really didn't mind because they were young and away from home. Pete was with me sometimes and didn't like it, but he didn't think he should create an incident over it. He was very protective.

We had to leave China as tensions were growing prior to World War II and all foreigners were told to leave. Mom and I left on a military transport and headed back to the United States. We returned to Washington, D.C. and our old house in South East DC. My dad followed at a later time. Pete and I were reunited when he was later stationed at Annapolis and came to DC to visit. We started dating and were married in July 1942. During the War he was stationed much of the time in the South Pacific.

I never thought that I would live again in the Philippines and was quite surprised when Pete got orders to go. I knew that it would be different after the War and it did look different, especially with all of the destruction the Philippines endured. It was also similar in that the people seemed the same and maybe even friendlier to Americans.

We were living in Hatboro, Pennsylvania when Pete received orders for Sangley

Point. We were hoping to go to Port Leote, Spain (also referred to as Port Lyautey, French Morocco) but for some reason didn't get those orders. We knew that each family had to secure reasonable housing off base to start and it was generally a 14 month wait to get base housing. Pete drove across country in September 1953 and left from Travis AFB to Tokyo on September 11, 1953. When Pete left I took Robert (Bob) and Jo Ann went to Arlington, Virginia to live with my parents until we were cleared to join him. We waited for months to hear when we would be able to join Pete. We wrote lots of letters (most of which I still have) and he said it was lonely without the family and how much he missed us. He would write and tell me about other wives and families who were on their way or already there. He told me all about what the place was like, the people and activities. He even tried to find the place I had lived in Manila.

USNS *Barret*

Finally we received our permission to join Pete. We were to take a military transport ship to the Philippine Island. It was the USNS *Barrett*. It was originally the SS *President Jackson* for the passenger service of American President Lines but was commandeered by the government and renamed in honor of Major General Charles D. Barrett USMC. It was operated by the Military Sea Transportation Service.

The *Barrett* was air-conditioned and the largest and nicest of the military transport ships. We were scheduled to depart on December 28, 1953 from San Francisco but actually set sail on December 30, 1953. I flew on American Airlines with Bob and Jo Ann a few days before we were scheduled to depart. Bob was so sick on the plane that they had to give him oxygen. It was not a pleasant trip.

Fortunately, the *Barrett* did turn out to be nice. Bob, Jo Ann and I had a cabin of our own because of Bob's age so we did not have to share with another family. There really was not a lot to do on board. We saw movies, ate and slept. There were fire drills to attend. There was a washer and no dryer on board. You were allowed to use an iron although not supposed to use it in your room. We had a very nice cabin boy, a Filipino man, who spoiled the children. Bob was sea sick a lot of the time and I would get him up early and get him up on deck for some fresh air. It took 18 days to reach Manila with stops at both Hawaii and Guam for a day each. We were so happy to finally arrive at Manila.

Pete had been working on getting us housing for months. He had been told it would be 14 months before we could get base housing (either new duplexes or a Quonset hut). That time got shorter as more people were transferred but we still were wait-listed when we arrived.

Pete had been arranging to move into a place the Villani family (he was a Navy pilot) lived in but that did not work out. He also tried to arrange to have a house built but there were issues with the pipes and having hot water.

We lived in two places off-base in Cavite. One was a compound of seven houses where we were able to rent a large house. The living quarters were upstairs. Underneath the living area was space for storage, a parking area for our automobile and nice living quarters for our housegirl. This was Certeza Court. There was a large water tower at the back of our house and the house faced the bay. The first place we lived was called Pig Alley and we had a two-story townhouse. This was a smaller compound (only three houses) but was closer to the base than Certeza Court.

We finally got permission to move onto the base in October 1954. I wasn't very happy about having to live in a Quonset hut but we didn't have a choice once we were on base. There were some newer duplexes but we did not get one of those. The Quonset had a great location right on Manila Bay. There was a corrugated fence I think to keep people from falling into the bay. There were lots of black sea urchins with shiny white diamond-like eyes in the water. The Quonset was larger than a typical Quonset. It had two bedrooms, a larger living and dining area, kitchen, large pantry and one bathroom. Outside there were banana trees and gardenia hedges which smelled rather fragrant. And of course there were lots of palm trees. There were also red ants and you learned to stay away from them. One time when Pete was gone, we had to pile all of the furniture in the middle of the living room because a typhoon was supposed to hit but luckily it did not. Another time we came back to the Quonset and when we walked into the kitchen there was a huge rat. I screamed and Pete had to calm me down. It took quite a long time before I felt comfortable in my kitchen again. I looked for that big rat for a long time.

Most of the military families and more well to do Filipino families had help — maids, and houseboys. Pilar, was our maid from the time we arrived until we left. She had worked for another family before us. She was in her early twenties and very friendly. She could have stayed with the young women who worked for the other families but she wanted to stay with us when we moved to the Quonset. We cleaned out the pantry and put a bed in there for her. It was very tiny, but she much preferred that to staying in the Quonset designated for the other young women who worked as maids. She never said why she didn't want to stay there. We also had a houseboy named Martin, who was Chinese, and worked for us while we lived off-base. Pilar did not like him for some reason.

Pilar helped around the house and sometimes looked after the kids. Pilar would play with the kids and take palm fronds and weave then into hats and animals. She was very creative. When she went to the market in Cavite she would frequently take Jo Ann with her. People there would ask if Jo Ann was her daughter, especially when they wore matching dresses. I had a dressmaker, Dahling, who was a great seamstress and made matching outfits for Jo Ann and me. I gave some of mine to Pilar and she liked wearing the same one Jo Ann did when they went to the market. I sometimes worried that someone might kidnap Jo Ann and I'd never see her again. Occasionally I also went to the market but usually I did my shopping at the Base commissary and Navy Exchange. We were able to get most of the same food and other things that we were able to buy in the States. We really did not eat much Filipino food.

We were only a few houses away from the Admiral's big house on base. Jo Ann became friends with the Admiral's wife and their dog, a big black standard poodle. She would go over to visit and play with the dog, who really liked to play hide and seek. They must have liked her because the Admiral's wife sometimes came to the Quonset and asked if Jo Ann could come and play with their dog! I was concerned that the dog might hurt her by accident because it was so big but he was always careful.

Both Bob and Jo Ann attended school on the base. It was basically a couple of Quonset huts. The school didn't have a real name until the1960's and then was named for John Paul Jones

One activity that I got involved in which I wasn't too thrilled about was learning to swim. I was really afraid of the water but was somehow convinced to take swimming lessons. Irma Kruzich, one of the squadron wives, was my instructor and she did not let me get away with anything. One time I stood at the edge of the pool and she wanted me to swim across to get my Red Cross certification. I was in tears and told her I couldn't do it but she insisted and I finally did it. Some how I was actually put in charge of teaching the little kids in the shallow end of the pool! I wonder how many of them ever got to the deep end!

There was a complete solar eclipse on June 20,1955. Manila was close to the center of it so we had great seats to view it. We were told not to look at the sun but someone made cardboard boxes with pinholes to look through and you could see the reflection in it. We were outside our Quonset and the kids were very excited as it started to grow darker. The kids thought it was great fun and went inside for a few minutes to play. The eclipse lasted for about 6 minutes and 1 second and until July 2009 when another total solar eclipse was recorded in China the one we saw had been the longest recorded one. It was great to have a chance to see a spectacular once-in-a-lifetime event and not even have to travel for it.

One day Pete was driving in Cavite and accidentally ran over a big sack of balut that a local woman had left in the road. Balut are fertilized duck or chicken eggs. The eggs are kept warm in the sun for about 8 days and then stored in baskets or bags to keep them warm. The balut is ready to eat in about 17 days. Pete said they smelled very bad, especially after he ran over them. He ended up paying for them and got the nickname of the Balut King. Of course, we never got the courage to try one!

Our squadron put on the play, ***The Shooting of Dan McGrew***. I don't remember who organized it or directed it but Pete was cast as Dan McGrew and I played a dancehall girl and wore a red satin sequined dress. The play was based on the poem of the same name written by Robert W. Service. I think Sangley was a strange setting for a play set in the late 1890's in the Yukon but everyone had a good time and seemed to enjoy themselves.

I kept busy with many activities. Besides playing bridge with the other wives, I also took bridge lessons and learned to play mah jongg. I also took hula and mambo lessons. I took Bob to various activities including Boy Scouts and Spanish classes. I took Jo Ann to ballet and Brownies. Both of them took art classes and swimming lessons. It seemed that I had activities scheduled daily.

The squadron had a lot of parties, either at someone's house or at the Officers Club. We all got along well. The wives played bridge and got together for lunches. We also took trips together.

There were a number of different trips that I took, either with the family or with friends.

We took a trip in April 1955 with the whole family to Baguio, which was up in the mountains. The drive was beautiful but very scary. The road was narrow and very winding. I'm sure I kept my eyes closed for a lot of the trip. At one place we saw a carabao in a field near the road so Pete and the kids got out and went over to pet it and take pictures. The kids really enjoyed it but it made me nervous. While in Baguio we stayed at Camp John Hay, a recreation area for the military. It was beautiful and much cooler there than at Sangley.

We went to a village where people still lived in the traditional manner. They lived in nipa huts which are thatched huts built on poles up off the ground. They dressed in traditional garments with the men wearing loin cloths and many of the women wore woven skirts. They made some beautiful wood carvings and statues of native people. We bought a number of them. We also heard that the people there killed and ate dogs. I'm not sure if that was true, but it was a bit shocking and hard to imagine killing animals that to Americans would be your favorite companion.

Some of the wives took trips to Hong Kong. We stayed in Kowloon and I remember we did a lot of shopping. Things seemed very inexpensive. You could have hand-sewn clothing made overnight. Before we arrived in Sangley, Pete had a suit made for his dress whites before his official ones arrived. He also had a Barong Tagalog made, which is the Filipino formal wear shirt (which was actually casual looking) made for him. President Magsaysay wore one to formal functions.

The squadron, wives, and children took a trip to Corregidor. It was interesting to see an historical place and walk around the site. You could see the gun embankments and tunnels and imagine what it might have been like to be one of the people defending it from the Japanese. Many people were sick because of some bad food (I think it was the octopus) from the night before when there was a party. Not a pleasant trip for many. I was feeling alright because I had not eaten any bad food.

During the time we were stationed at Sangley the squadron made many flights. I worried when Pete was flying. The wives never knew where their husbands were flying but we knew it could be dangerous. When our husbands were gone the rest of the squadron members made sure to look after the wives of the men who were flying and all of the wives took care of each other.

We left the Philippines on September 26. 1955 aboard the USNS D.I. *Sulton* and headed for our next station at North Island, San Diego. As a family we had lots of wonderful memories of our time in the Philippines. It was sad to leave really good friends but we knew some of our friends would be in San Diego when we arrived. We also knew that we would all share a great bond which would keep us involved in each others' lives in the future. I don't think that we realized just how strong a bond was forged during that time or that we would still be close after fifty years. It was a very special time and a very special group of people and an experience for which I will always be grateful.

U. S. N. S.

GENERAL D. I. SULTAN

Lee Chapola, left, and husband Pete at right.

Eleanor (Lee) Chapola

Autobiography

February 28, 2011

I was born in Washington, D.C. and lived there until age 16 when I left for the Philippines and then China with my parents. Returning to the States, I went back to school and then went to work for the Navy at the Navy Yard in Washington.

Pete and I continued to get to know one another during this time. After Pete and I married, he was stationed in the South Pacific for part of World War II. Upon his return, we were stationed at a number of places around the country. I kept busy raising our three children and volunteering my time at various organizatons.

We were living in Virginia when Pete retired from the service in 1960. He continued to work as a marketing consultant for several firms and then retired in 1979. We lived in Virginia until we moved to Colorado in 2003. It was hard to leave Virginia, which I considered home for over forty years and where my heart still remains.

Joan Angier

TRIP TO THE PHILIPPINES AND JAPAN

It had almost seemed too easy. There I was seven months pregnant, standing on the Navy pier in Honolulu clutching the hand of my nineteen month old daughter, Ann, and staring at the MSTS *Maxwell*, the Navy transport that was about to take me on the final leg of my "Great Adventure."

I had just said good-bye to my new best friend, Helen Bohnhoff, who had brought us down to the pier. I had known her for only three months, but she had played "Mother Hen" to my "Babe in the Woods" knowledge of what the future had in store for me.

We had lived in Pensacola where Don began flight training and then moved to Kansas for him to complete that training. Thinking back to that snowy, February day in Kansas, I remembered Don's voice on the phone saying he had orders to the Philippine Islands, NAS Sangley Point to be exact. He was to be in a squadron called VW-3 Detachment A. Then he added that he (we) would go to Hawaii first as he had to go to a six week All Weather Flight Course at Barbers Point. He was saying many other things, but I think my mind went blank at the words Philippine Islands, Hawaii. Growing up I had traveled extensively with my parents but only in the United States and Canada. I was so excited that I could hardly wait for him to come home and tell me the rest of the news. Then I came down to earth. How would I tell my parents? I was an only child. Ann was their only grandchild, and there was the fact I was pregnant. Don's father was not well and while Don had two brothers that would not make it any easier.

When Don came home, we talked about some of the details of our impending journey. We did call our parents. They were very subdued, but happy that we were thrilled with the orders.

Don received his wings shortly after that. We were to sail from San Francisco the first week in April on the MSTS *Daniel Sulton*, a military transport. Our two-week home visit to Burlington, Iowa was poignant, but no one uttered the words, "Why don't you stay here and have the baby?"

We arrived in San Francisco early, as we had to get passports, physicals, series of shots, fill out paperwork necessary for the Philippine government, and obtain a Visa before entering that country.

Then two days before we were to sail Ann got sick. We called the hotel doctor and after examining her, he could not find anything that would delay our trip. There was a military doctor on board the ship, he said, so if anything else developed he assured us that the ship doctor could take care of it. Little did we know she had measles!

I remember the ride to the docks in San Francisco and all of the things we had to do before we could board. It seemed to take forever, but once we were finally in our cabin, reality set in. I was so lucky that Don was with me at least for the trip to Hawaii.

The first few days we encountered some ground swells. Most everyone stayed in their cabins including Don and Ann — except me. I was in the dining room at every meal and ate everything including the fried eggs for breakfast. Maybe I should have been the sailor in the family!

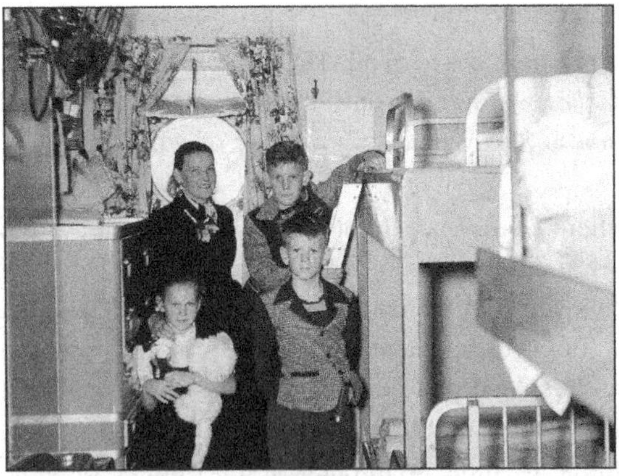

Accommodations for families of junior officers were spartan aboard *General Daniel I. Sultan* (T-AP-120) This is the one assigned to the family of USAF Captain Maher in their cabin. Mothers were warned to keep children away from portholes as they were dangerous.

The weather began to change. The air smelled differently. I never understood the word balmy until I arrived in Hawaii. A friend we knew from Pensacola met us and took us to her house in Wainai. It was right on the water and fortunately, for us there was another house for rent nearby. The fact that it was unfurnished did not matter.

Don bought a "used" car to drive back and forth to the Base. The class had already started so he had to wait a couple of weeks for the next one. What wonderful days those were. Playing in the sand with Ann, swimming in the warm tropical ocean, eating the delicious tropical food, admiring the beautiful flowers and birds, were many of the activities we enjoyed.

The first thing I had bought upon arrival was a maternity swimsuit. We spent most of our days at the beach only a few miles up the road. I had no furniture except for a card table and two chairs, and a mattress on the floor for sleeping. (Ann slept in her playpen). My housekeeping consisted of opening the front door to sweep out the sand. In the little town of Nanakuli, about five miles away, there was an outdoor movie. We sat on those hard benches with Annie in Don's lap and thought it was heaven. Besides, the drug store in Nanakuli made the best chocolate sodas I had ever eaten.

The Navy informed Don that since I was so far along in my pregnancy, I would not be able to travel until after the baby was born. Six weeks after the birth! He would have to leave in July. That was a reality check as he only had two more weeks of classes. Helen came to the rescue. She said Ann and I could move in with her. She had three boys and a sister-in-law who already lived with her — her husband was away on deployment. That was the way it was.

Tripler Army Hospital, Oahu, Hawaii

A few days later, I had an appointment with an O.B. doctor at Tripler Hospital. As we talked, I mentioned my hometown, Burlington, Iowa. He looked at me rather strangely. He told me that was where he was from before he went into the Army. His family lived a few blocks from Don's family. We knew his younger brother, but did not know about his older brother who was in the military. I started to tell him why I was going to have to stay in Hawaii to have the baby. He said, (and I remember his exact words) "No, you don't have to do that. I'll sign your release form to travel on to the Philippines." I could not believe it, but that is precisely what happened.

Don had to leave the first part of July and fly to the Philippines to find housing, but now here I was at the pier waiting to board ship. It had all happened to quickly.

Finally, it was time to go on-board. A sailor took pity on me and helped me with my luggage, including the playpen. Annie and I were lucky. We had a cabin to ourselves. I did not know anyone, but all of the other wives and children seemed to be confused, too. The ship was also carrying troops to the Asian theater.

The new Chaplain and his wife and daughter were organizing things for the children. It included a few hours of day care for the younger ones. It allowed the mother a couple of hours to go to a movie, play bingo, or just nap! I remember my biggest worry. My cabin had a porthole that needed to be open. The little fan in the room did not help much in the warm tropical weather. Ann was a climber and I was afraid she would fall out of that porthole!

Four of us on the ship were pregnant. The doctor was an OB/GYN specialist. His last assignment was on a troop transport and he was delighted with this new duty. Every morning he would ask each one of us how we were feeling.

Our ship docked at Guam to drop off some of the passengers and a few of the troops. Shortly after we tied up to our berth there was a knock on my cabin door. Don was standing there! He said, "Aren't you ready to get off the ship?" I just looked at him. "But, we're going on to the Philippines, right?" He just laughed. He had flown over to meet the ship and spend the day with us. He had arranged a tour of the island in an old Navy bus. Lucky for me, I did not have the baby after bumping around for an hour and a half. Finally it was time to re-board. It was so good to see Don and to know that in a few more days we would be in Manila.

School in Quonset Huts — NAS Sangley Point, RP.

Annie had been so good on this trip. She was extremely shy when she was little. She did not fuss and seemed to enjoy the playtime with the other little ones. She would sleep with me at times but with my stomach seeming to grow by the hour, that lower bunk was crowded!

The big day finally arrived. We docked in Manila around mid morning. Sure enough, Don was there to meet us. Our car had arrived earlier and he had driven around the Bay from NAS Sangley Point to pick us up. It took a while to collect everything and say my good-byes to some of the friends I had made. Later, Bev Woodward, one of the other pregnant wives would live only three houses from us in Certeza Court. On the drive from Manila to Cavite the sights, sounds, and smells of this new country overwhelmed me. It was a great feeling of the three of us together again.

Less than six weeks after I arrived in the Philippines the fourth member of our family put in his appearance, John Roland Angier.

Local school in Cavite City.

In June, the squadron was commissioned VQ-1. It was no longer a detachment of VW-3. This change also brought a change in location. Our home Base would now be NAS Iwakuni, Japan.

The trip from the tropics of the Philippines was a big contrast to the September arrival in Japan. The kiddos and I were lucky there was a military flight leaving from Manila. I was getting special transportation because I was a High School teacher. Ann and John were so good on the flight — except for a little accident of a cup of cocoa spilled all over my dress. Fortunately, the colors nearly matched.

The wife of the Base Commanding Officer greeted us as we landed at NAS Iwakuni, Japan. After collecting our luggage, she drove us to her home where she served us lemonade and cookies. She seemed overly pleased that I was finally there.

The problem, though, began as I realized high school consisted of nine kids spread over all four years. They were anxiously awaiting my arrival. The next thing I realized was I had a bare room, 15 desks and no books or supplies.

The next couple of days were spent with the education officer, Milt Hutchinson, trying to scrounge up "make do" material. We took a boat to an Army Base for that!

In the meantime, Don stopped by my classroom where I was pondering what to do. When he knocked at the door and I saw who it was, the tears rolled down my cheeks. The problem seemed insurmountable. How was I to teach nine children the entire curriculum required for four years of high school? By the next day, I knew that I had to face reality. Don had found two Japanese women to cook, clean, and take care of Ann and John.

Margaret Burdg, my neighbor who turned out to be a great friend, gave me her welcome insight. She was still recovering from the death of her two-day-old son. Margaret was also a high school teacher so we talked about how to proceed with courses for these children and things seemed to be brighter. The three girls in the group seemed to be more interested in the cute sailors and marines stationed on the base than school. I explained to them that school came first and social activities were secondary. A couple of the older students were a great help and I think we were all learning.

I wished I had had more free time to explore Iwakuni and its places of interest. Teaching took a lot of my time. That fall we had an early snow — the first I'd seen in a long time.

The elementary school and the high school were in the same building and shared the same Principal. She came to work everyday with her hair up in rollers so she would be ready for the Officer's Club social hours in the evening!

While I had little free time, we as a family, did take a long weekend trip or two visiting as much of the local area as we could. We did not remain in Japan long enough to see the cherry trees in bloom. However, we did enjoy the trips around the countryside and experience the friendly Japanese people. They showed no animosity even though the war had been over for only a little more than ten years. Iwakuni is an hour's trip from Hiroshima and crews were busy with pick and shovel rebuilding that city.

Don's orders State Side came in early January and we were excited. The trip back was not as comfortable as the trip over, but Don was with us, and that made up for so much. Being able to stop for two days in San Francisco made up for the rest.

I was amazed that so much had changed in the United States over that time period. That was the year all the cars were so colorful, not just gray or tan.

However, I did not realize how laid back, and yes, easy our life had been. I would not, could not, trade that experience for anything.

Japanese fishing village on the Inland Sea.

Joan Angier

Autobiography

April 22, 2011

She died November 2, 2012.

I was born, raised, educated, and married in Iowa. My husband, Don, was one of the "Shadow Warriors" who piloted the P4M-1Q Navy planes from bases in the Philippines and Japan. During Don's tour there, our family was able to live on these bases.

I was needed to teach at the dependents' schools on both these bases; sixth grade in the Philippines and ALL high school subjects in Japan. Securing high school textbooks required transporting them in a small launch from another base to ours!

Fortunately, live-in help was available to care for our two small children while Dan flew and I taught. During the rest of Don's Navy career, I did more dependents' school teaching, and especially enjoyed it in Spain.

After retirement, we moved back to Iowa. Our daughter is a special ed. Teacher; our son is an airline pilot. Don has passed away. I keep active enjoying family and hobbies.

Alice Sumner

FERRET FLYERS

I did not care for the word ferret as describing our Squadron's occupation. However, I knew that in some way I was going to have to embrace that word so I consulted my dictionary. The following is the definition that appears in my Britannica World Language Dictionary belonging to a set of Encyclopedia:

Ferret (fer'it) n. 1. A small, red-eyed domesticated polecat of Europe, about 14 inches long; used to hunt rodents and other vermin. 2. A black-footed weasel of the western United States, which preys on prairie dogs. –adj. Ferret-like; red, like the eyes of a ferret. 1. To drive out of hiding with a ferret. 2. To search out by careful investigation; without. 3. To hunt with ferrets. 4. To hunt by means of ferrets. 5. To search.

It became clear to me that the word was used correctly in the sense of being an adjective meaning "to search out by careful investigation: without" . . . "to search". And therefore, I found the correct meaning to describe the Squadron's "secret flying missions".

Knowing the proper meaning gave more depth of understanding to the gravity of importance of their missions. We were all very privileged to have been a part of this undertaking.

Respectfully submitted,
Alice H. Sumner

Martin P4M-1Q Mercator, Fleet Air Reconnaissance Squadron One
MCAS Iwakuni, Japan, 1957

Philippine Islands — Here We Come!!!
The Family of a Ferret Flyer!!!!!

It was in the Fall of 1953 when Don walked in the door and said, "Honey, I've got my orders". We had been expecting reassignment after his two years as Flight Instructor in the Pensacola, Florida area. "Looks like we're going to California." That sounded exciting to me, especially when he said "the San Diego area". Wow, we had to get the ball rolling! I was thrilled to think we were going to the West Coast as I had never visited that area. But first, the we I am referring to, were Lieutenant Donald A. Sumner, USN, myself, Alice, Bill who had just turned four and Nancy at 19 months of age. I was 25 years old and I was expecting our third child the end of the year. We had to sell our little home and make the long trip to the coast.

It didn't take very long to sell our house. It was quite small, very modest, no cabinet doors or closet doors, just about as plain and simple as could be and also very inexpensive. Our monthly mortgage payments were only about $32.00. There were two bedrooms with closet alcoves, a tiny bath with an improvised shower, a living room and small kitchen, all with black asphalt tile floors. There was a shed in back and a fair sized yard. No frills, but it was "our home".

We packed the car with our luggage in a way that we could put down a blanket or two to have a level area so that the two children could either lie down or play quietly with toys or books while traveling. That was in the days before mandatory car seats. And so, we drove, a happy family with many adventures ahead. In due time we arrived in San Diego, found a motel and settled in while Don found out about his orders. What we found out was he had "further orders" to Naval Air Station (NAS) Sangley Point in the Philippine Islands to a top-secret squadron VW-3, Det. A. Dependents could not travel along with the military before that time, and my pregnancy did not allow me to travel overseas until after the birth of our baby. Wow, news we didn't expect to hear; however, Donald was very excited at the prospect of being a part of this very elite squadron, which was later renamed

VQ-1. But the kids and I didn't have a home!!! What to do, with the further complication of a new baby yet to be born. We were very depressed at the prospect of being separated and had to come up with plans to take us through the next several months ahead. It would have been easiest for the children and me to just stay in the San Diego area, get an apartment and wait out the time element, but the prospect of our new baby to be and the children had to be carefully considered. We finally decided it would be safest for all concerned to drive the children and me back to Chicago, Illinois where there was family to help us through this time. Having made that decision, difficult as it was, we decided to lighten up a little and visit the marvelous San Diego Zoo, which was a perfect antidote at that time. The children loved it and we were able to relax and begin to make some plans as to how the next few months would be lived.

Now remember, we had lived in Florida, a semi-tropical climate, and were now headed North with winter just around the corner. Don's parents had the most room for us. So that is where the kids and I were to stay. We had to buy some winter clothing as it was cold and snowy. Don left the car with me and he flew back to San Diego to embark on his next duty station. We spent a lonely Thanksgiving time that year, although it was pleasant being around family we hadn't seen in awhile.

My in-laws' home, where we were staying, was in River Forest, a suburb of Chicago, while my parents lived on the North Side of Chicago. I would have preferred to stay with Dad and Mom but they lived in an apartment and there really was no extra space for us at that time. My in-laws turned over their large bedroom for Bill, Nancy and myself, which was very nice. I rented a crib for Nancy and Bill slept on a rollaway bed. Bill had a very favorite "Smokey Bear", a fairly large stuffed animal which he dragged with him everywhere. I think it was his best friend and to this day he still has it. It helped him survive the stress of being separated. And then we waited. The days seemed to drag by. Each Tuesday I would drive to Chicago with the kids to visit and spend part of the day with Mom. She was an Angel. We always had fun together, baking cookies or having tea parties, special times. So very enjoyable. The weather was cold and nasty and it was not easy to take the kids out to play and time seemed to stand still.

It was necessary for me to line up a civilian obstetrician for prenatal care, and also visit the nearest military facility to receive all necessary inoculations from diseases prevalent in the Far East. Glenview Naval Station was the nearest military installation and was quite a few miles away. We set up a schedule to receive our shots and at 8 and 9 months pregnant I did a lot of driving back and forth on icy and slippery roads to complete our retinue of typhus, typhoid, tetanus, and cholera inoculations plus smallpox vaccinations. My children were not too happy to have to go through this with me, but it was very necessary. My dad worried about my driving in the icy weather on slippery roads in my "condition" and so he came to our rescue of his own accord several visits as I neared the end of my pregnancy.

Dad didn't want me driving on the icy roads. He always was and always will be "my hero", or one of the Angels who would touch my life at this time.

To keep busy, there were doctor appointments and the long treks to the Base for our shots and trying to keep my children happy. It was not as pleasant being at my in-laws home as everything was kept in perfect order and poor Bill, being a four year old was very restless and didn't get along too well with his Grandmother and what I kept hearing was "Butch" isn't allowed to do this or that or whatever. Butch was Don's younger brother, about 13 at the time. Bill would misbehave and I would hear. "You must do something about this child, he can't get away with (whatever it was he was doing). Consequently, Bill was a nervous wreck and I was even more of a wreck knowing that we were making their life miserable. And, so we waited for the baby to be born.

My due date was about Christmas, but Christmas came and went. My Angel, Dad, would swing by as a surprise once in a awhile to take the kids and me out for an ice cream cone or treat. I guess he sensed how tense I was, out of my comfort zone, away from my husband, waiting for a baby, getting shots, seeing my obstetrician, caring for my children and waiting to make plans for a long trip to my new home in the Philippine Islands eventually. I finally asked Dad if he and Mom could please have Bill stay with them while I finished waiting for my baby. Dad and Mom were delighted and I was much relieved to know he was with them, and so, we waited.

Finally my Dr. decided to prescribe castor oil to help nudge this baby into the World. And so, the evening of January 8, 1954 it was necessary for me to ask my father-in-law to please take me to the hospital, that I was going to have the baby. Mercifully the delivery went well and beautiful red-haired Barbara Anne was born at 8 lbs. 2 ½ oz., 21 inches long. While I was in the hospital I asked my Mom and Dad if I could please somehow go back to their place and stay with them as I wasn't at all happy at my in-laws. They were just not enjoying having us there. I was grateful that they had provided room for us a little over two months but I couldn't go back there because it made me so nervous. My parents were so understanding and in their usual way of unconditional love somehow made room for us. We were surprised with beautiful bouquets of flowers, all with a single red rose in the center twice a week for a month. Donald had somehow arranged for the delivery and prepayment before he left us. Each bouquet delivery brought a tear of thankfulness and appreciation for this reminder of his love for us. It was a beautiful thing for him to have done and so very dearly appreciated.

Nancy stayed with the Sumners for a week after I got out of the hospital but I wanted her with me at my parents' apartment. We set up a crib for Nancy in my brother's bedroom, bless his precious heart, and Bill slept on the sofa and I slept on a roll away cot in the dining room. Needless to say, it was very cozy, but as my Mom always would say, "we'll manage", and we did. I had a bassinette for Barbara and had diaper service for two because Nancy was not yet trained. We had ropes strung up in the kitchen as we had so few clothes with us I would wash up everything in the bathtub in the evening and they would be dry by morning. We managed just fine that way. My parents only had laundry privileges in the basement of their building on one day a week, but we were able to get along okay in this manner.

Finally, our checklist was coming together. Necessary shots received, baby born, now it was time to get our passport in order and a photo taken. At three weeks of age, I wrapped Barbara up in many warm blankets, and Bill and Nancy in their snowsuits and I had a borrowed coat and we ventured out in sub zero Chicago nasty weather to get our passport photo. It was a composite of me seated with Barbara held up on my lap and Bill and Nancy on either side of me. It worked!!! The authorities allowed it and I was issued our passport for four. Now, to await the letter stating when I was booked for transportation to the Philippine Islands.

Mother and Dad usually spent a month or so in Florida during the cold winter months and they graciously took Bill with them. Dad was so concerned about Bill's being so nervous and distraught and it was the perfect antidote for him to be with Mom and Dad and playing on the beach. They enjoyed him so, and he had a wonderful time with my precious parents. So, for the next six weeks my adult brother, Edward who lived with my parents, and I held down the fort, so to speak, and time went by. Nancy turned two in March 1 and I invited several of the relatives and my in-laws for cake and ice cream. Nancy was adorable and when we sang Happy Birthday to her, she put up her finger and said "Shush", she was so embarrassed.

I hoped to hear about my transportation sometime in March, but we weren't scheduled until April. I needed to get my car out to the West Coast to be shipped to the Philippines. We found someone to drive it to San Francisco and I arranged for a cousin of mine to drive it to Oakland to be shipped to the Philippines. Things were falling into place. I was feeling well, Bill was completely restored and a happy little boy again. Nancy a sweet two year old, and Barbara was thriving and a very good baby. We were ready!!! Finally we had our departure date to travel aboard the USNS General *M. M. Patrick*. The children and I flew out of Chicago with all of the relatives at the airport to bid us farewell on April 10, 1954. It was the last time I was to see Don's favorite Grandmother, Toto. She adored all of us and we thought the world and all of her. My mother and Dad had tears streaming down their faces as they knew it would be quite awhile until we returned to the United States. I choked back the tears but I was on my way to my husband and the children's daddy and so I was very happy.

When we landed in San Francisco, it was dark outside and I had my hands full with my three children, purse, diaper bag, formula and portfolio of travel documents and shot records. I had found a strap in a specialty store, which encircled Barbara and me and helped hold her to me with at least free one arm. I had Nancy in a special child's harness so I could hold the strap and not lose her and Bill was my big boy and stayed by my side with his "Smokey Bear" in tow. We needed to pick up our luggage, find a cab and get to our hotel. Suddenly, Angels appeared again and someone took my baggage stubs from my hand and said, "We'll get your bags and take you to wherever you are staying!!! Please allow us to introduce ourselves; we are Dr. and Mrs. Brown." I guess they could see that I had no one meeting us, and they were so gracious and kind, and they got our very meager luggage and brought us to the Colonial Hotel. I hope I adequately thanked them, because that was such a generous and lovely thing for them to do. I remember them in my prayers. We had a comfortable room and the hotel kitchen allowed me to make up Barbara's formula and store the bottles in their refrigerator. There was a playroom and Bill could run off some steam. We settled in for the next few days of more adventures!!! My luggage was filled with diapers, canned formula, and the very few clothes we had with us. Back to the old routine of washing out our clothes at night and hanging them on ropes to be worn again the next day.

I needed to get to Oakland to sign necessary papers to ship the automobile. The hotel found a baby sitter for the girls, and Bill and I received directions to go by bus to the Ship Yard to take care of that. All went well, but when we went to return to our hotel, we were waiting at the bus stop for quite a long time. An automobile drove up and a gentleman opened his door and informed me that the last bus had left for the day. He offered to show me his credentials and help us get to where we needed to go!!! Another Angel!! He truly was a gentleman and brought us directly to our hotel. I guess I must have a sign over my head which says "Help" or something because this was extraordinary!!! I was so fortunate. Nancy was happy to have me back and the dear little lady who had been watching the girls had read to Nancy and played with her, and watched over Barbara.

USNS General *M. M. Patrick* leaving San Francisco harbor. Photo USN.

Now, the next step was to check in at the Federal Building to make sure our passport was okay and get our final instructions. In the morning Bill, Nancy, Barbara and I had a taxi come to pick us up and drive us to the Federal Building. We did all necessary checking in and received our debarkation time notification, and early the next morning we left the hotel and boarded another taxi with our luggage to take us to the ship, the USNS *General M. M. Patrick*, (the illustrious Mickey Mouse Patrick), which was to be our home for the next three weeks. I did not know anyone aboard the ship. Our room had no porthole, was way below sea level (troop quarters, I believe) and very, very warm. No air conditioning!! The children developed terrible cases of prickly heat rash and were miserable with that. There were two bunks and a crib in between. Bill was happy with an upper bunk, Nancy in the lower bunk, Barbara in the crib, and I had the other lower bunk. Lo and behold another lady appeared! This was her room too. I don't even remember her name but do know that she was a new bride on her way to Guam, which was our only stop on the way to the Philippines. She was the most considerate lady ever, making herself very scarce during the day, only sleeping and showering. I sometimes wonder if she ever had any children, after having been thrown in as a roommate with me and my family. The kids were really good though. We shared a bathroom with the group next door to us, a total of five woman and four children. Somehow, it all worked out! I was almost relieved that Nancy was still in diapers along with Barbara!!!

I remember the thrill of the ship disembarking on April 13, 1954 and the feel of the rocking of the boat. Quite a sensation and I learned to love it!! Within an hour after we left port, whistles blew and the order from a loud speaker said, 'all hands on deck'. We were to report topside with our life jackets on. This was a safety drill. I had Barbara strapped in with me, Nancy bound up in hers and Bill in his and we made our way up the ladders. The kids were crying, thoroughly confused, I was trying to hold on to Nancy so I wouldn't lose her and keep an eye on Bill. I remember sailing under the Golden Gate Bridge and thinking it was the very most beautiful sight!!! It was all mightily confusing and noisy and very strange for everyone, to say the least, but we managed to get through this portion of our journey and we were on our way!!!

The children and I were assigned the first serving for our meal settings. This meant 7:00 a.m. for breakfast, 11:00 a.m. for lunch and 4:00 p.m. for dinner. I had Barbara strapped to me, Nancy in her harness and Bill by my side. I don't think I ate very much because I had to cut up food for the kids and see to it that they were fed and we were only allowed 15 minutes until we had to clear out for the next group. Unreal!!! I was thin to begin with but lost over 10 pounds on that trip. Didn't have enough time to eat and it was so hot! And then, it was going to get really crazy!!! They were sanding or scraping the decks immediately above our room and the racket screaming overhead was almost unbearable. This went on for about a week. Nancy would just sit on her bed and the tears would roll down her face. I would try to read to her and play with her while Barbara slept. Bill found a group of kids to play with in an organized setting so he kept occupied. We only left the room for meals and I would check on Bill playing with the group of children so it was pretty confining. We passed through the International Time Zone and received special certificates for having done so. That was exciting! I don't remember going on deck except during our life vest drill. In the evenings, after the children were in bed and sleeping some of the other mothers and I would sit on the deck in the hallways outside of our rooms and we'd talk and it was so friendly and nice. One evening, one of the ladies offered to keep watch over my family so that I could go on deck. There was a movie showing but I was not interested in that. What I saw when I got to the open air upper deck was our majestic galaxy, slowly rocking back and forth as the ship was sailing over very calm waters. And, up above I viewed the Southern Cross seen only in that hemisphere. Never have I seen anything so beautiful! It seemed as though it was a very good and comforting omen. I only stayed on the upper deck for about a half hour, but it was so calming and peaceful to be under the rocking heavens with the Southern Cross overhead! I didn't mind returning to the area which the children and I now called "home". We were very near the kitchen area so some adventuresome souls would find a tray with cake or some other goodies and we'd share and have fun that way in the evenings. There were washers and dryers aboard so I could keep up with laundry that way. But, there were still two in diapers. They sold disposable diapers aboard the ship, flat rectangular sheets of absorbent something and plastic. I needed to use two at a time to go around Nancy and then put a regular diaper over that to hold it all together and plastic pants. Not the best arrangement but it was innovative and worked. I didn't have to worry about strange toilets for a little girl. Nancy was trained shortly after we arrived at NAS Sangley Point.

We sailed into the harbor at Manila Bay on May 4, 1954 and I was shocked to see the remains of the many battleships and rusting hulks of tankers at that time. They were the remains of huge battles and had not yet been removed under the

These are the Japanese ships that we Americans sank in Manila Bay. The Bay is muddy and shallow and these ships sat in the mud. Since these pictures were taken the Japanese have raised and salvaged all 750 sunken ships.

reparation rules of war. The Philippine fishermen made good use of the ships and used them as makeshift shelters and for fishing. Many buildings in Manila had been bombed and there were ruins everywhere. To say the least it was very eerie and served as a reminder as to what the military had to do at that time to win the war against the Japanese.

We finally docked and I got the children ready and I remember walking down the gangway ladder with the three children when I saw Donald. It was a wonderful reunion. The children were so happy to see their Daddy, and there was a new little girl to greet!!! It had been almost six months since we had seen each other. I do remember saying, "Here, take a couple of kids"!!! The burden had been heavy, but immediately forgotten upon our family rejoining!!! After going through this experience alone, I knew that I had grown immensely and would be able to handle anything that life would hand me!!! Little did I know that I would be tested many more times in the future? I had built up a strong 'inner core'. I was to tap on that resource many times in the future! Thank heavens for Angels!!

Pil-Am-Life Court — Cavite City

Then on to Cavite City, Philippines. This was an hour's drive through the local barrios or towns, definitely a Third World Country. There were ruins everywhere and I was going to see the big carabaos soaking in the ditches along the way. They were water buffalos, "beasts of burden" and helped work the fields. Finally, we drove through the Cavite City Checkpoint where we were scrutinized carefully and then not too much further we were in our new home called Phil-Am-Life Court. Our unit was two stories in the fashion of today's townhouses. We were one of the first ones and being an end unit, we had more windows, several of which overlooked the Island of Corregidor. It was so soon after the war had ended that you could feel the 'ghosts' of the action which had taken place there and of the fine men and women who had led our country to victory.

Don had arranged for a housegirl, Pina to help with the house, which was the local custom. She was a wonderful help to me. Our quarters had a living room, kitchen with eating space, hot running water, and three bedrooms upstairs, one for Don and me and we had Barbara in with us for a while and Bill and Nancy shared a bedroom and Pina had the other small one. This worked for a while, but then we decided that the girls should be together and Bill have his own room and Pina stayed at a local nearby home. I think she preferred being with us, but we needed

A view of the neighbors in Cavite City

the room for the children. We were quite comfortable as we had all our household effects with us. Everything had been sent over in big wooden boxes and the packers had hammered nails into the furniture while crating everything, but it didn't really matter as it was our home again. The closets were "hot lockers", or closets with neon fluorescent tubes on the floor to heat them to keep down the humidity. There was no air conditioning. Don's big leather boots turned green with ½ inch of mildew and I was constantly airing things in the sun to cut down on the growth.

NAS Sangley Point was situated near Cavite City, Philippine Islands. Our squadron VW-3 Det A was stationed there. The planes were very impressive large multi-engine aircraft painted navy/black and had no insignia. Their mission was highly classified, and I never knew what Don was doing. After having flown many missions, he became a Patrol Plane Commander. Their flights were secret. I would never know if he would be gone overnight, a week or three weeks, sometimes as long as a month, if they were flying to the United States for some purpose. I would usually know then, but otherwise, they were never able to tell their wives anything about the duties of the Squadron. One time he did show me the weapon he carried and was ordered to use to self-annihilate if ever captured. I didn't ask any questions. He couldn't have divulged any secret plans.

But, so life in the P.I. began!! We were so happy to be a family again but were in for a very different way of life. To begin with, I had our car for transportation and Don had a motor scooter to get back and forth to the Base. After a year, we sold the car and I had to use the Jeepneys for my transportation.

These were ordinary WWII discarded military jeeps which had been put into service by the natives and were used as 'sort of taxis'. This was quite an experience as it put me in very close proximity with the locals, which included pigs and chickens as well as stinking fishing nets. Once I had Nancy with me and the people would touch her chubby arms and say "Beebee". Their children were very tiny and petite and Nancy was a typical curvy healthy American child and they were in awe!!! Don tried to teach me to drive the motor scooter but one lesson was enough for me! Scared me half to death!!! However, we had a lot of fun with it as it had a second seat in back of the driver. I could and did sit sidesaddle holding on to Don, many times with Bill standing in front of Don and Nancy in back of me on the luggage rack. We were young and could do anything!!! Barbara would be safely at home with the housegirl. This was, of course, before the days of safety helmets and that

A Jeepney is used as a bus, taxi and to transport most anything, including animals.

type of gear. We felt invincible, but we really weren't as Don was to learn one morning on his way to work. He was scooting along on his regular way to work when a carabao decided to argue as to who had the 'right of way'. Guess who won? It wasn't Don and he ended up in the muddy ditch and got a very infuriating dunking. He came back to our quarters fuming up a storm and had to not only wash and dry off his motor scooter to clean up the mud but strip off all his clothing because he was mud from head to toe. Thank heavens for house girls because they cleaned up all those messy clothes and shoes. After showering and changing clothes and settling down, it was back to work again for him.

Our quarters were very near the checkpoint and sometimes at night, we would hear gunshots. I guess they were trying to keep out the Huks, (Hukbalahaps) who were the communists in the Philippines, from trying to take over the country. Our particular compound was fairly large and surrounded by a high concrete wall which had large broken shards of glass imbedded along the top. One day, a little girl living in the compound caught her arm on the glass while jumping and playing with the other children. She really injured her arm and tore muscles. Needless to say, her parents rushed her to the Base for emergency medical care and I do believe the Medics airlifted her to Clark AFB, which had a more sophisticated hospital structure. She had a long recovery and needed much physical therapy. At night, we were guarded by an elderly Philippine gentleman named Mateo. He carried a long rifle and walked back and forth patrolling our area. Don had given him some clothing, which he no longer wore, and there was a sweater that Mateo particularly loved. He wore it every night because I suppose the night dampness would cause it to be chilly if you were outdoors. He became our devotee and many a night when Don was gone on a special flight, I would see Mateo sitting on our front steps with his rifle over his knees guarding the area. I always felt safe.

Cavite City

I will never forget the dirt roads with deep trenches alongside for drainage, and the carabao wallowing in the mud and the little children wearing only t-shirts to cover their chests to ward off TB, and no underpants as they freely used the open outdoors to urinate and defecate. Many were of mixed parentage, probably the unknown children of American GIs who had been stationed there and had liaisons with the pretty young women.

Caraboa cooling off in ditch full of water next to the road

The local natives lived in Nipa huts, which were elevated on stakes. Their animals, chickens, pigs, and goats lived underneath. Peeking into one of the huts I saw a bed, which was nothing more than bedsprings. There was no mattress or coverings of any type. They thought they were very fortunate to have a bed. I remember how treasured an empty Crisco can was to the house girls and used for their cooking or storage purposes. There was always the unforgettable smell of cooking fish heads and rice, which was their diet, if they were lucky.

The local marketplace was an open-air area which sold fruits, vegetables, meat and materials. There was no electricity or refrigeration where carcasses of animals hung on hooks, covered with insects flying around. The Military families did not buy meat anywhere but the Commissary. Even so, we were cautioned to soak all of our fruits and vegetables, even potatoes which were to be peeled, in a solution of Clorox and water to kill germs. I bought beautiful mangos and found a recipe in my cookbook for mango ice cream which I thought would be a very welcome treat. The fruit was quite beautiful, peachy tangerine color, oval in shape and about the size of the palm of your hand. I was tempted to eat one, so after the recommended cleaning procedures with Clorox and water, I sliced into the succulent fruit, tasting rather like a peach. In the middle was a large almond shaped pit. I devoured the entire mango fruit sans the pit. I loved it!!! However, unbeknownst to me I was very allergic to mango and my face swelled like a balloon, my lips were swollen quite huge and I could hardly see out of my eyes. So, a quick ride to the Naval Station Dispensary for treatment. My cure was to be shots of Benadryl at 8:00 in the morning and 8:00 at night until the swelling subsided. The swelling lasted about a week and I was a very sorry sight. My eyes were swollen at an angle and Don said I looked like a blonde Chinese. It was hard to breathe. However, I learned!!! Never again have I come near mangos, but I still think they are very beautiful!!! But, not for me!!!

The marketplace sold pretty fabrics very inexpensively and dressmakers were readily available. For two to four pesos you could have a simple tropical wear dress tailored. I had clothing made for the girls and myself very reasonably. There was not very much to be had in the local Exchange. If Don went on a flight to the States he would bring back dresses for me to wear and also clothes for the children. Shoes were a different matter. I was unable to keep up with the growth of their feet and as a result of wearing shoes that were improperly fitted Nancy developed a hammertoe. I have felt very badly about this ever since, but the children were not to go barefooted at any time because of the danger of picking up roundworm from the soil.

One of the local traditions was balut and beer. This was an extreme delicacy the Filipinos really enjoyed. They would bury eggs which had an embryo duck or chicken inside for a month or so. They would ferment and when the time was right they would be cracked open and downed with beer. A tradition we never were tempted to try but a true Filipino cherished.

The Commissary did not sell fresh milk. The children finally became accustomed to drinking powdered milk mixed with boiled water and cooled. Often I would add chocolate syrup just to make drinking the mix more palatable. Other than that, I believe we could get most of the basics necessary for a healthy diet from the Commissary. Over time I guess we lost the taste of milk and accepted the necessity of drinking powdered milk. It was such a treat to drink fresh milk again once we returned to the States!

Life in Phil-Am-Life-Court was mostly peaceful. Our quarters were pleasant and airy, but there were ants and roaches to contend with, being that we were in the Tropics. There were quite a few children for Bill and Nancy to play with. The Philippine American Life Insurance Company had erected these housing units to help the military accommodate families of the officers and men attached to our Squadron. There would be occasional family oriented parties at Halloween, Christmas and Easter. So it was pleasant living there. One Halloween we dressed Nancy as a sheriff with a stocking over her face. I had painted a mustache on her face and had to cut a hole in the stocking so she could suck her thumb. She was dressed in outgrown cowboy gear of Bill's and was riding on a little rocking horse. She was the STAR and won a little bamboo chair as First Prize. Such fun!! Bill was Casper the Ghost. I had sewn his costume, and he also received a prize. Nancy still has the little bamboo chair in her home. Each of her children has enjoyed it as much as she did. There was a little girl with cerebral palsy, who used a wheelchair, and her mother had decorated it with flowers and she received a prize as PRETTIEST. She was so thrilled. At Christmas, there would be a Santa Claus, at Easter, egg hunts and so the ambience was pleasant!

There were Red Cross Swimming Lessons available for the wives, so I thought it would be a good idea to learn further how to swim correctly. I had problems learning how to do the crawl, but was able to do better with the Red Cross method, but it was not until many years later that I summed up the fact that I had my face too deeply in the water to be able to turn my head to adequately catch another breath of air. However, I managed to learn to sidestroke and backstroke and made an attempt at the breaststroke. That turned out to be difficult for me, but I was very happy with what I had learned. The instructor also was giving Junior and Senior Life Saving and Survival Classes. I struggled with these, but not being able to dive made it a bit difficult for me. However, I did receive several certificates of my achievements and I was proud of that. I would rush home after the swimming lessons to fix lunch for the children and for Donald who always came home for lunch, not having time to change from bathing suit to other suitable clothing and Don nicknamed me Chlorine La Poole!!! Then it was time for the "accomplished" swimmers to help with swimming lessons for the children. To this day I do not dive. I assisted with beginners but my own children didn't learn to swim until later on when we were stationed in Monterey, California.

It wasn't easy to get everything taken care of with three children, but it was very nice to have the assistance of the house girls. Pina was with us for the better part of a year when her Mother committed suicide by ingesting rat poison. Pina returned to her northern province and sent her sister Lourdes to help me, but she didn't help very much. She did not have the sweet personality that Pina had and was not trained. We missed Pina very dearly. We were then to go through a succession of girls. Conchita, a nice enough girl but with BO so bad it would almost knock you over. Then there was another very sweet girl, Rosalie, who turned out to have TB and we all had to be X-rayed and tested to see if we were contaminated. Another thought Barbara was a boy, and was she surprised to discover Barbara was indeed a little girl. That maid had never worked for Americans and didn't know how to do the simplest things, so it was easier not to have her. For a while I got along with a *lavendera*, one who does the laundry and ironing. She was a very quiet, hard working older woman I liked a lot. I was able to manage the rest, but it was nice to have a maid as they would clean the floors and polish everything to a shine. They swept with a small broom and used half coconut husks and put wax on the floors and then did a hip hop dance with a foot on the husk and sashaying back and forth. Later when the Gillooly's returned to the States, I had their maids, Letty and Dominga, and they were very nicely trained.

During the time we still had Pina with us, we took a trip to Bagiuo in the mountains. Admiral Spruance had resided in this beautiful mountain environment with cool, fresh breezes. It was very enjoyable to be away from Cavite and the Station. The trip was a challenge because driving up the mountain consisted of about 32 hairpin turns, but it was worth it when we arrived. It was pleasant, much

Switchback road up the mountain to Baguio

cooler and breezy. We could see for miles. Pina took the opportunity to visit her family and we were together with just our family. Nancy had come down with impetigo, a skin infection from the soil, which leaves some nasty sores, and so I had to bring her for medical attention daily but we still enjoyed ourselves because it was such an escape. Every evening we would gather with the other vacationers in a dining hall for our dinner. This was always very lovely, with an organist for entertainment. When he would see our family enter he would strike up the song, "How much is that Doggy in the Window". Bill and Nancy loved that, and Barbara was old enough to catch the excitement. While in Bagiuo we were able to see the Ting-a-kling dance performed. The performers used long bamboo poles which they strike together in a rhythm, and balance candles on their hands as they danced between the poles. We were all so enthralled with that and the children tried to emulate the dance with long sticks back in the compound but that resulted in bruised ankles, so we decided we would just remember what we had seen. The Ting-a-kling dance was ever so graceful and synchronized and beautifully done, truly an art form!!!

We also visited Taal Crater, which believe it or not is an island in a lake in an island in a lake (Taal) in an island (Luzon) in the sea. Taal is a volcano, 50 km. south of Manila and blew out in 1911. It was a very fascinating sight to behold. We also drove by the Philippine equivalent of one of our Military Academies where they schooled and trained young Filipino men who met the high standards.

It was lovely, cool and breezy in the mountains but basically the P.I. was hot and muggy, being tropical. Our feet would swell and there was much prickly heat amongst the children plus the impetigo requiring extreme care when tending to any sores. Also, there was the problem of children picking up worm infestations from the soil. One little girl, the daughter of an enlisted man who worked with Donald actually was so acutely infested with worms she choked and smothered to death. This problem could be caused by airborne contamination as well as come through contact with the soil.

We often had rain and it would really inundate everything. One time we were literally swamped from a typhoon. Our quarters stood up under the stress quite well, however the roof leaked and water rushed in, not on the second floor, but between the floors and soaked everything. I had every pot and pan, bowl, cup, etc. spread out trying to catch the water leaking in. Quite an experience! There was a storage space where I kept linens under a window on the second floor and water soaked through the side of the house into the linens until we discovered the deluge. It was a full time job for all of us keeping the receptacles emptied and replaced until the typhoon passed. We were grateful there were no injuries and we soon dried up but had to watch carefully for mildew.

Upon occasion, several wives would join up and we would take a boat to Manila to visit Pierre who gave us our haircuts. We would do a little shopping and return on the boat. I remember one Sunday that Donald was gone, another wife,

Jean Hanson and her little girl and myself with my children decided to just board the boat and take a ride to Manila in order to spend a few hours. We planned to return on the same boat. Our boat captain was not too happy with us when we didn't get off in Manila because it would have given him an excuse to visit a bar or whatever. We were just happy to spend part of a lonely day away from the base on the cooler, breezy sea.

There was a Filipino custom of St. John the Baptist's Day. The local people would take buckets of water and soak as many people as they could; I guess their form of baptism. Unfortunately one of the wives was on the station, all dressed up to have a prenatal check when she was soaked thoroughly by one of these people. There was nothing she could do but keep her appointment with her doctor, although she was quite upset having to appear soaking wet, but it was "their" way. They meant no harm, just their tradition.

We used Military Payment Certificates (MPC) for money when on Base and Filipino money, Pesos and Centavos in town. There was a special building at the entrance to NAS Sangley Point where money was exchanged for that purpose. This never posed much of a problem until I was installed as Treasurer of the Station Wives Club and when it was time to pay for the Luncheons we would have, keeping a record of Pesos, Centavos and Military Script. It proved to be a bit of a challenge!!!

We had parties, almost monthly. A much-needed relief for our Squadron because of the nature of their flying and responsibilities. A lovely party could be catered for about $80.00. This would include liquor and hors d'oeuvres, but in particular their specialty, lumpia. We loved lumpia which were small egg roll wrappers stuffed with cabbage and shrimp and several dips. We all loved them and devoured many. They were so special!!!

There was a change of command that was heralded by much Navy tradition, brass band and all. As many of us as could, attended and witnessed the "ruffles and flourishes" of an official military tradition. I remember the pomp and circumstance. It was a fabulous part of Navy Life which is unparalleled in the civilian capacity. I remember having a new dress made and being so proud.

Pauly Varner and I hosted a special luncheon for Pink Sweeten, wife of the Commanding Officer 'Hap' Sweeten, who was leaving. We wanted everything to be very special and decided on a menu of everything pink. I don't remember the menu, but rest assured it was as pink as we could get it. We also asked every wife who was attending to wear something pink, the more the better. The, punch was pink also, but I don't quite remember what was in it, but it was wonderful!!! There was some left over so we brought it home with us to sip as we tallied up the luncheon receipts and balanced the books. Needless to say we had a super wonderful time putting the finishing touches on a very special Pink Luncheon!!!

Living off Base was quite an experience as it put us in close touch with the native people, and we could see much first hand. The drive to NAS Sangley Point was mostly down dirt roads, which passed by an area called Bennie Boy Alley. We were cautioned to be very careful going through that particular stretch of road, so we were! And, I was to learn why!!! Donald was gone on a mission and so I was alone this one evening. My neighbors asked if I would like to come with them to see a parade. They didn't plan to be gone long. That sounded nice and the children were being watched over so I went with them. We lined up alongside the road and witnessed the parade of floats of the most beautiful women dressed in native pina cloth formal dress with big butterfly sleeves. Everyone was gorgeous! I was amazed when my friends informed me that they were all men who were homosexual. We didn't even use the word 'gay' in those days. I had never heard of anything like this, even at age 25. Later when Donald returned from his long flight he explained this to me and I was aghast. These people were so beautiful, lovely and dressed so elegantly!!! I kept on learning!!!

Trip on the *Margaret* to Corregidor

I learned a few words in Tagalog which was the Philippine language, at least in the region where we lived. The words were 'maganda balaklak', or pretty flower. There was lovely tropical foliage everywhere. We also had lizards, little ones of course, and their purpose was to eliminate some of the insects. We welcomed them, the lizards that is, and they even lived in our quarters with us. The children delighted in chasing them and sometimes the tails would fall off, however they would grow back again. They were fascinating little creatures doing a very helpful job.

A group of officers and their families sailed on a boat to visit Corregidor. This was quite an experience. We brought Bill along as we thought he was old enough to remember the trip. He was almost six years old. What we found was, of course, very eerie, having been the site of violent battle. It had been General McArthur's headquarters and everything was now overgrown, rusted and falling apart. We took a short walk into one of the tunnels but it was so spooky, you could feel the atrocious memories of conflict surrounding you. Needless to say, we didn't stay very long but it was quite incredible to see. We could see Corregidor from our quarters and from a distance it looked like a very peaceful island, but upon close scrutiny there was no doubt as to the part it had played in history. We were all glad to get back aboard the boat and return to our 'home'.

Several families were returning to the States so that meant that Base Housing

Aerial view of NAS Sangley Point

would be available for some of us living in Cavite City. We were on the list and really needed three bedrooms but only two bedroom units were available. However, the quarters were quite spacious with more than adequate room for us. We were able to have the three children share one bedroom with sufficient space and privacy for each. So, we were to say ‘goodbye’ to the dusty roads and deep ditches and the checkpoint, rice fields and salt flats that had been our neighborhood for almost a year. Onward to quarters on the Base!!!

Naval Air Station Sangley Point, Philippine Islands

Sangley Point was located on a strip of land about a mile long that jutted out into Manila Bay. This land was mostly runway and also contained the Naval Station and housing. There was a school and church located across the street from one another. They were both Quonset Huts. The Hospital/Dispensary was nearby but was made of cement blocks up on pylons. There were frangipani trees everywhere, so pretty with their orange blossoms, also bougainvillea and large oak trees. The Commissary, Navy Exchange and Snack Bar were a block away. Nothing was air-conditioned. The Squadron was located nearby next to the long runway. The duties of Squadron VW3-Det. A were top secret. The planes were very impressive, P4M-1Q, blue/black in color with no insignia Our quarters were rather near the runway and the sounds of the aircraft taking off and landing would be very noticeable.

So, life aboard the Station began. The quarters, duplexes with large windows and shutters and a very large screened porch, which was to be a multi-use room, mostly for the two youngest as a playroom was very functional. I was very happy there. Bill could be outdoors in the large playground area in back of the quarters playing on the equipment and with friends. Nancy was watched over more closely nearby and Barbara was allowed to explore the enclosed screened area. One of the maids caught a bug, like a cicada, and tied a string around it and hung it from the ceiling allowing it to fly around, yet still controlled. The kids got such a kick out of that. We had a nice roomy kitchen and a very sufficient dining area and living room. There was always a breeze so we were quite comfortable. Upstairs we had a very large bedroom with huge windows that were shuttered and a large bathroom, and the children's very nicely sized bedroom. We were close to all the amenities and Don surprised me with a bicycle. I loved riding it. One day I had Nancy on the back rack, holding onto me and off we went for a nice ride. We went thru the Gate and were riding in the area outside the Base near a Filipino school. All of a sudden Nancy cried out and I stopped. She had gotten her foot caught in the spokes and her shoe was ripped off and her foot was bleeding. Suddenly a Filipino man grabbed her and ran off with her to a nearby building with me following, making sure to turn the wheel of my bike and lock it so it wouldn't be stolen and ran to catch up with them. The man was a teacher in the school and he was in the office with Nancy smearing Vaseline, from a huge jar, all over her foot. Their method of First Aid. I wasn't so sure! I thanked the man for being so kind. I was able to quiet Nancy down and put her back on the bicycle, this time on the seat, and holding on to the handlebars with me holding her and walking beside her, pushing the bicycle until we were back aboard the Base and to the Hospital/Dispensary area. That must have been about a mile walk. The Doctors were quite concerned with her injured foot and they took X-rays. They were afraid the skin would slough off requiring a skin graft. However, they cleansed the area, shaking their heads over the Vaseline and properly medicated her and wrapped a goodly sized bandage on her foot with directions to not walk on that foot, and make an appointment for a return visit. In due time and after much toting of my daughter, her foot healed beautifully. Perhaps the Vaseline treatment was not quite right it was done with such a caring manner from a complete stranger . . . yet another Angel! Pretty soon she was back to being her own very peppy, sweet, happy self.

P.I. President's Palace "Malacanyang"

We were alerted to be very careful and not use any beach areas nor wade or play in streams because of the danger of schistosomiasis. This was a worm parasite that had its origin in the mountain areas where infected people would urinate in the streams and contaminated the water as it traveled downstream. The organism infested the person, their abdomens would swell and they would waste away. There was no known cure.The Station Wives Club organized a visit to the Malacañang Palace. Magsaysay was president of the Philippines at the time. The Palace was decorated very lovely and with beautiful gardens. The Guest House was located in the same area immediately behind the Palace. It was small but also beautifully landscaped.

We visited the Santiago Prison and grounds where prisoners from the war were tortured. On the walls and ceiling you could see the fingernail scrapings made by the prisoners in their agony.

During this time we were told that there would be a complete eclipse of the sun. The hour came and everyone gathered outdoors to watch. We were instructed not to look directly at the sun except through indirect methods. It was truly a phenomenon. The moon began passing in front of the sun and it became very cool and dark, almost greenish in color. The birds ceased singing and tucked their heads under their wings as if it were nighttime. It actually became chilly and dark as the sun's rays were blocked. Gradually the moon passed and the skies brightened with the sun's rays and the birds once again began to sing.

Bill and Nancy attended Sunday School once we moved aboard the base and he enjoyed that. Nancy had other ideas. Her teacher said she couldn't keep her in class as Nancy would rather be outside swinging on the swings. In late June school was to begin and Bill would be a first grader. He was ready!!!

Margaret Burdg and I planned a trip to Hong Kong. Her husband had been there recently and he wanted Margaret to visit also. Donald was not scheduled for any flights so it was possible for me to go as there would be a parent available for the children. It was not a good idea for husbands and wives to go together at the same time, it being a fact that we were so far from the United States in case of a catastrophic happened of whatever type. Margaret was expecting so we planned carefully. We flew Philippine Air Lines and planned to stay three days. Everything went well. The landings and takeoffs at the airport are very tricky! In Hong Kong the approach was between canyon walls with a very sharp turn before the landing. The wingtips almost touch the walls of the canyon. We kept our eyes closed as we took off and landed. The rate of exchange was 5/1. The American dollar was worth the equivalent of five Hong Kong dollars. Our days were planned for selecting materials, having clothes made, fittings, shopping, and then we would return to our hotel and rest for a while, have tea, and shop a bit more. We dined at a lovely restaurant in our hotel and enjoyed watching the elegant local people being very festive. One day we hired a barefooted coolie to carry us by rickshaw to Ada Lum's very unique jewelry shop. We asked him to wait but he didn't understand that and he raised a ruckus and had the local police come after us to pay him. The misunderstanding was smoothed over and we paid him well and he returned us to our hotel. I look back at that experience and realize we were very lucky to have come out of it unscathed. We didn't buy very much jewelry as I remember, because of the disturbance. We also visited the Tiger Balm Gardens which were quite beautiful and unusual. An entrepreneur had invented a special cream which was to cure all aches and pains and made a lot of money. He created the Gardens as his gift of 'thank you'. Don Burdg had made friends with the tailors who made our clothing and they invited Margaret and me to join them to see Kowloon in the evening. We took a cable car to the top of Mount Victoria. The cable cars were so constructed as to fit the slope of the mountain. The seats were level but the cars were on a steep slant. It was a very beautiful sight to see the cities of Hong Kong and Kowloon and the ships docked below all lit up and reflected in the sea between. And, then it was time to return to NAS Sangley Point. I missed my family.

In the early evenings I would walk with the children, Barbara in her stroller, to where the tradition of Taps would be held at the flag lowering. This was very impressive with the bugle sounding as the flag was lowered and folded. It never ceased to send goose bumps up and down my spine. Our beautiful flag!!!

There were times when Don and his crew would come home very early in the morning, tired and wound up from a long surveillance flight, open up cold San Miguel beers, and have a few to relax and unwind. The children would usually be awake and they loved to be a part of watching this. Nancy named Gary Grau the 'Giggly Man' because he had a way of making everything fun. The fellows would relax for a while and then all go home to catch up on some sleep. And this was our life in the Philippine Islands.

We had parties at the Officer's Club and sometimes hosted visiting dignitaries from other countries. It was lovely to see women dressed in saris and very formal dress. We were also entertained by Colonel Latif, a military attache' who performed a dance in complete native costume. I believe he was from Thailand. All were once in a lifetime experiences!

We were to see more changes! In June 1955 the Squadron was commissioned VQ-1, and the homeport was changed from NAS Sangley Point to NAS Iwakuni, Japan. We were going to move again! This was a little more complicated with Bill in school, beginning in June, but he learned to adjust. We were scheduled to be in Japan only six weeks in order for the experienced Officers to supervise the new personnel and get the Squadron into an efficient operating mode. Donald, Stretch Varner and Jim Edixon had further orders to the Post Graduate Meteorology School in Monterey, California.

Bill had about four months in school in the Philippines and then we were to move. This did not go too smoothly. We had a departure date which was delayed. Our possessions were packed and on and their way to the States. My children and I had only a few pots, pans, and minimum clothing with us to wait it out. In the meantime Kathy, our new Commanding Officer's wife and five very unruly kids were assigned to stay with me, which was utter chaos for a two-bedroom house. She complained loudly and was assigned an empty three-bedroom house temporarily. Finally, after about a week's delay we got an all clear and we were to board the aircraft. I had new outfits sewn for the girls, navy blue jump suits and they looked so sweet. They each had a special toy to carry along and a book to read for the flight. Bill was wearing his coonskin cap Don had brought back from a trip to the U.S. Davy Crockett was all the rage at that time. Bill wore it everywhere. He too carried a book to read. I was dressed in skirt and blouse, ala Navy tradition. We finally took off, but after a short time, the pilot turned back for some reason. We were sitting in bucket seats along the length of the aircraft, and luggage was piled and secured in the center. The problem was rectified after a long and unpleasant wait and we finally took off again.

It was a very nostalgic time leaving the Philippines, inasmuch as it had been our home base, giving us a sense of security at a time when security was at a premium. I have never again been with such a wonderful group of people. The Officers were all outstanding, dedicated, extremely devoted to their duties, and the wonderful, wonderful wives were such a delight to know as we shared our lives.

The experience had been the worst of times and the best of times in so many ways. We made the best of our hardships and had joys beyond measure. It was a time of togetherness when the World had been torn apart and was still fraught with unknown danger. We had played a big part in keeping our United States safe with the secret electronic countermeasure missions which were flown by our heroic military. I am so proud of the achievements of our Squadron. And, to this day the flights and missions continue. I will never forget the friendships we made and the reunions so many years later.

Everyone was hungry and thirsty and box lunches were passed out amongst us. There was fried chicken and other food. Pretty soon some very unruly children began throwing food and the chicken bones all over the plane. My children didn't participate in this because they knew that when Mom said 'no' she meant it. This went on for a while until someone established order. It was getting late and was already dark. Finally, we arrived at our destination only to be told we couldn't land and had to circle the airport until the problem was fixed. So, we circled and circled and circled, my three children became very nauseated, and all three began throwing up all over me and themselves. We mopped up as well as we could but now we smelled to high heaven. This is the way we greeted Donald at our new duty station, NAS Iwakuni, Japan!!!

Naval Air Station Iwakuni, Japan

It was late and dark when we arrived. We were taken somewhere we could wash up and get our luggage to hunt for warmer clothing as it was now Fall and we were much further North than the Philippine Islands. It was so good to see Donald as he was able to help me through the rest of this journey. It was not over yet! We boarded busses and drove for quite a while to where we boarded rickety, covered boats which then transported us to Nijimura where Army Housing awaited us. This was quite a long ride and everyone was so tired. Upon reaching Nijimura, we were met by more busses and taken to our assigned quarters. I believe it was about 4:00 in the morning at this time so everyone was exhausted. Upon first glance the Housing didn't look so bad. They were barracks-like buildings and furnished simply, but adequately. At this point all we wanted to do was curl up in a bed and get some much-needed sleep. It had been a very long and frustrating journey. The Army Wives had tried to make us feel welcome with a few staples in the refrigerator so in the morning I was able to get breakfast for us. The kids were starving and we were all rather reeling from the experiences of the day before, but we were ready for more adventure. First of all were the Japanese pots and pans. There was an assortment of strange looking cookery, everything with handles that were about three feet long. We deduced that they were probably used over open fires in Japan, but manipulating them on a stove was a learning process. I remember fixing scrambled eggs in a frying pan that was so strange; however, we were so hungry all we wanted to do was eat. The toaster was a very archaic thing which burned the toast and yet it tasted good. We found the peanut butter and jelly and all was well with the world.

I was to find out that the Commissary was about four blocks away and we had no transportation. It is not an easy matter to carry heavy groceries for a family of five that distance. I sent out an S.O.S. to Donald and he brought me a brand new bright red wagon and that is how I hauled groceries for the time we were living in Nijimura. This worked pretty well unless the kids wanted to get in the wagon with the groceries and have me pull everything. I might have made an allowance for Barbara but I soon learned to ask the housegirl to watch the girls while I shopped. Our Japanese maid's name was Matsui and she was very sweet and I liked her a lot. Bill was enrolled in his second first grade while in Japan because their classes began in August, so he sort of just fit right in. Life in Japan had begun!!!

Don continued to fly his missions plus reorganizing the Squadron in their new spaces. The Officers living in Nijimura would leave for NAS Iwakuni aboard busses and then transfer to boats very early each morning and back home to their families in the evening. They never saw the light of day where we lived except on the weekends. It was an adjustment we all had to make.

We saw the Japanese carry 'honey buckets' on yokes on their shoulders to fertilize their fields of rice and vegetables. This was human and animal excrement. The vegetables grew to monstrous sizes. We had to be careful what we bought in the way of our food. It was safest to use canned. The Commissary proved to be adequate, however very minimal with displays set up in cardboard boxes instead of shelves. We were able to manage a balanced diet; however, there still was no fresh milk available.

A group of the Squadron people gathered one evening to enjoy sukiyaki, a Japanese delicacy. This was served by Geisha girls, who were elegant and lovely in their native kimonos and elaborate hairdos. They cooked and danced for us. The table was low to the floor and we sat on pillows and ate with chopsticks. I became very adept at handling the chopsticks and taught my children to eat their rice with them. That was fun. We learned to love Oriental cooking.

Horse transport in Japan

The Army Wives extended their hospitality to us in a very cordial manner, inviting us to luncheons and bridge. I am not an avid bridge player but agreed to play. They assigned a partner to me, a very buxom English woman with glasses that pinched on her nose, no less. She was a rather formidable looking lady. I know why they assigned me, no one else wanted to play as her partner! So the bidding began, I thought she said "hearts" in her accent so I said "two hearts". Well, what she had said was "pass", so when I put down my cards she almost had a heart attack because it was rather a disaster and I received looks that would KILL. Oh well, we were going to be there only six weeks. I never played cards there again. Once was enough!

We visited Hiroshima where the atomic bomb had been dropped. It was utter devastation except for the one building which partially survived. Ruins were everywhere and it was very humbling to all of us. So many lives had been lost in World War II. (I was to return many years later to visit my youngest son John and his family at Christmas time, as he was a Marine Captain stationed at MCAS Iwakuni. I traveled alone from Tampa, Florida to Osaka, Japan, transferred to a bullet train for about another 300 mile ride and then to a taxi. While visiting my family at Christmas, we took a trip to Hiroshima and I was astounded at the changes. There were telephone poles and wires and people in Western dress and cell phones and cars and buildings all over, a very busy city. The one partial building that had survived the atomic blast was still standing as a memorial to those who had died).

Another weekend we traveled to Kyoto and visited the beautiful temples and gardens, all in traditional Japanese elegance. It was quite interesting to see. We had the children with us and we brought Matsui along. Donald was not feeling too well with a heavy cold so he stayed at the hotel. The children stayed with Matsui for a

Hiroshima in 1954. The building to the left rear of the memorial is known as the Atomic Bomb Building or Genbaku Dome. The first operational atomic bomb was supposed to target the Aioi Bridge nearby, but detonated almost directly above the dome of the building.

few hours and I found a taxi and traveled to a very interesting shopping area. I just showed the address on the brochure and so there was no language barrier. Their taxis were interesting as the cars drive on the left side of the roads so the driver sat on the right side.

Another nearby trip which I made with several of the wives was to see a local parade of the school children. I had my children with me. I was very blonde and my children were blonde, light brunette, and a redhead. We watched the school children and they were dressed in their native black pants and blue jackets. They all looked alike as they had black hair, bobbed and with bangs. It was interesting to see. There were a group of Australian soldiers nearby also watching. I had the misfortune of a Japanese man coming up to me and hitting me on my arm and saying "bo okara". It frightened me and I gathered my children close. I was ready to call out to the Australian soldiers to help if there was anything more, but we left soon after. I told Matsui about the incident and when I said "bo okara" she became very upset and said, "Bad, very bad". I have never found out the meaning of the word, but it was not a good thing that is for sure. I was very happy to return to our quarters. I didn't venture out anywhere but the Commissary after that and very soon it was time to pack up one more time and return to the States.

Pauly and Stretch Varner, and Donald, myself and Bill, Nancy and Barbara were to leave Nijimura early in the morning. We got to the boat landing to learn that the coxswain would be delayed due to the after effects of a hangover. The crash boat could only operate on one of the two engines. Someone went to get the coxswain and someone else radioed the Base. The coxswain was told to drive the boat on the one good engine at slow speed, and another crash boat was being sent to meet us somewhere in the middle of the Inland Sea. There we transferred from one boat to the other, very carefully handing over the three children and luggage. Following Navy Wives tradition, Pauly and I were dressed in proper skirts and nylons. It was quite an experience handing over the luggage, and the children, but mostly ourselves from one boat to the other in the middle of the Inland Sea in straight skirts. Amidst much laughter, we climbed over the three-foot high boat railings that the sailors were holding tightly together. The new crash boat sped at full speed to the Base where the plane waited to take us to Tokyo. Just before landing at Kadena, the Tokyo Airport, we could see beautiful Mt. Fuji. From the airport, we traveled on to Yokohama where our ship to the States was docked.

It was unbelievable, but we were going to travel aboard the USNS *General M. M. Patrick* one more time!!! Only this time Donald was with us to share the experience. We had the same type of stateroom as when I traveled to the Philippine Islands with the exception that this time we had a porthole!!! There were two bunks and a crib in the center. Bill and Don had the top bunks and Nancy and I had the bottom bunks. Barbara was in the crib. And, so we sailed in December. This was the North Pacific Ocean. The seas were quite rough at that time of year. We went through a terrible, very frightening storm. We were taking 33 degree lists, side to side, front to back, over and over again. Looking outside the porthole all you could see was grey . . . the ocean, the skies, the ship. It was a very violent storm that lasted several days. We had to tie the crib in place as it was flying from one end of the stateroom to the other. Nancy stayed in her bunk. They couldn't put food on the tables The chairs were falling over in the dining room They did set out crackers left in their boxes so there was something to nibble on but everyone was very nauseated and throwing up and it was awful!! Eventually the storm subsided but the seas were still somewhat rough. We docked in Adak, Alaska to pick up passengers returning to the States. We all gathered on the decks dressed as warmly as we could as it was very cold and we saw the beauty of Alaska. Even in winter!!! White on white on white, the majestic mountains in the background. We didn't stay docked very long and resumed sailing and finally reached Seattle, our stateside port. When we landed, I was very emotional and went down on my knees and kissed our Earth. How good it was to be back in our beloved United States. From there we flew to San Francisco, where we stayed a couple of days to pick up the car Donald had ordered when we were in Japan, and then on to Monterey, California, and Naval Post Graduate School for our next tour of duty.

Addendum

I wish to thank all he people in my life who made it possible for me to join my husband and the Squadron in the Philippine Islands and Japan, half a world away, to share his ties with the most wonderful people in the world. First of all, my parents, my in-laws, my Angels who appeared so many times to lend a helping hand in ways that were so unexpected and deeply appreciated, and then there were and are the members of our Squadron. Together we were 'quite a bunch', and shared a camaraderie unequalled before or since. They are Bud and Ruby Abbott, Don (dec) and Joan Angier, Don and Margaret Burdg, Pete (dec) and Lee Chapola, Glen (dec) and Marilyn Carlstrom, Jim and Mary Jane Edixon, Dottie Flood, whose husband Frank was lost on a surveillance mission, Jack and Ursula Gillooly, Bob Galloway, Gary and Nancy Grau, Bob (dec.) and Jean Hanson, Andy and Irma Kruzich, Barney (dec) and Jinny Martin, Karle Naggs, Dean and LaVon (dec) North, Jack and DaLoris Reynolds, Stretch (dec) and Pauly Varner, and the many others I have failed to mention.

We were so far away from our families and the States but somehow we made our own Family and deep friendships evolved from that fact. My own family grew and remained quite healthy, albeit some of the unavoidable local irritants. I guess we were blessed by that Southern Cross constellation shining upon us overhead.

We have, over the years, enjoyed memorable reunions which took us back to the time we shared when we were a Ferret Family in the Philippines. It was as though time stood still . . .

Forever in my Memory,
Alice H. Sumner

Alice Sumner

Don Sumner

Alice Sumner

Autobiography

December 14, 2011

I was born on Memorial Day, May 30, 1928 as the parade was passing by the hospital, in Oak Park, Illinois. My parents had emigrated from Holland, my mother at age 15 with her parents and brother and 3 sisters. My father came to this country at age 20 with $120.00 in his pocket to find work. My parents met and married in River Forest, Illinois. I had one brother, Edward. We lived in Belleview, Illinois but when the Great Depression hit, my parents lost all their savings and home. We were forced to live with grandparents and other relatives in River Forest, Illinois. Later we moved to River Grove, Illinois where we lived with my grandparents until the economy stabilized. Then we moved to Chicago, Illinois and I attended school from seventh grade trough high school. I was the first in the family to graduate from high school.

I worked as a secretary for Union Switch and Signal Co. in downtown Chicago on Michigan Blvd. across from the Art Institute, overlooking Lake Michigan for two years before I was married to my high school sweetheart, Donald in Corpus Christi, Texas on August 6, 1948. Our son, William Ashton was born in Chicago, Illinois on June 29, 1949 and we moved to Patuxent River, Md. We lived in a small cabin in Tall Timbers, Md. Next move was to Pensacola, Florida and daughter Nancy Alida was born on March 1, 1952. Barbara Anne was born January 8, 1954 but that is included in my story.

We next lived in Cavite City, Philippine Islands, and then in Iwakuni, Japan until being transferred to Monterey California where Thomas Edward was born on December 2, 1956. Then on to Jacksonville, Florida for shipboard duty and then the Hurricane Hunters in Roosevelt Roads, Puerto Rico, where John David was born on May 26, 1961. On to Washington, D.C., Alexandria, Virginia and later Eden Prairie, Minnesota, then Largo, Florida where I took care of my Mother, and worked for the Dept. of Social Services as an Administrative Assistant until I retired at age 70. I live near the Gulf of Mexico and enjoy when my family and friends come to visit. I was privileged to always be a stay-at-home, full-time mother and homemaker. My hobbies are gourmet cooking, and I won a blue ribbon at the Minnesota State Fair for my weaving. I also sewed, knitted, crocheted, canned fruits and vegetables, made quilts and volunteered at the hospital. My five children have grown to be fabulous, accomplished, loving, kind adults. My nine grandchildren are very special people and I claim two great grandchildren. My family is the World to me!!! I loved being a Military Wife and am happy being retired and loving living!!!

Pauly Varner

THOSE FARAWAY PLACES WITH THE STRANGE-SOUNDING NAMES

The year was 1953, and my Navy pilot husband's orders read, "Proceed to San Diego, CA for further assignment". A short time later our suitcases were packed and in the car. The movers had our household goods packed. One shipment was going to storage, one to San Diego, and a security shipment to my parents in Massachusetts. The shipment to San Diego was just one wooden box of equipment needed for living while awaiting the rest of our belongings from storage. The security shipment was our silverware.

It was sad to say farewell to our little home in Pensacola, Florida, but a more profound and exciting time was in store for us. This house was our first big purchase in our young married life. The total cost of the house was $7400.00. After the initial payment, our monthly payments were a whopping $45.00 per month.

We decided to make a visit to the relatives in Massachusetts before heading west. We had no idea where our next duty station would be. Our friends and relatives in Massachusetts were anxious to know where we would be going. Stretch had completed the six-month photography school before leaving Pensacola and a photo reconnaissance squadron was a possibility. All we knew is that it would be somewhere in the western hemisphere. We even entertained the idea of going to Hawaii. Little did we know that it would be the Philippines?

I was delighted when Stretch told me that he was going to Naval Air Station (NAS) Sangley Point, Philippine Islands (P.I.), but then he told me that I could not go with him. Everyone called him Stretch, but his given name was Eldon. He was a Jr., and his father never called him by his real name. He called him Sonny when younger, but in high school the basketball coach called him Stretch. He was 6"4" and the name stuck. Stretch explained that housing for dependents was limited and we had to wait until a place was available before I could join him. I drove him to Travis Air Force Base (AFB), and we said our sad good-byes on a beautiful California day in October.

I was excited about going to a foreign country. I knew it was a tropical climate, and thoughts of beautiful beaches, palm trees, and gorgeous flowers came to mind. I had visions of shopping for clothes to wear in a tropical climate and then remembered it was October and that that type of clothing was not available in stores. Oh well, and I thought, I did pack my featherweight sewing machine in the wooden box that was being sent to San Diego.

The holiday season was approaching, and I did not want to drive back to Massachusetts to await my letter telling me that housing was available and giving instructions on how to proceed to the Philippines. My parents were furious when I told them I was renting an apartment in San Jose until I could leave for the Philippines. Earlier my father had told me that he did not think my marriage would last. Now he speculated that Stretch was abandoning me by leaving me to fend for myself. I was excited about going to a foreign country and that thought never entered my mind. My friends in Massachusetts used to sing me the song, "Far Away Places" when I told them I was going to marry a Navy pilot and travel to many faraway countries. Now it was coming true.

I had friends in the San Francisco area and the time passed quickly. Shopping for material and patterns for clothing suitable for the tropical climate, buying Christmas gifts and shipping them to the relatives, and spending time with my friends kept me busy.

Finally, I received the letter authorizing me to go to the Philippines. February was the big month I was to report to the Federal Building in San Francisco. February 6, 1954 was the date listed for me to sail on the USNS *Barrett*. I was busy obtaining a passport and visa in addition to getting all the required shots. I drove my car to the Oakland Navy Yard for shipment to the Philippines and then took a bus back to San Jose. The owners of the apartment complex where I lived treated me like a daughter and the landlord would take me shopping. His wife did not drive and before I took my car to Oakland, I would take her shopping.

One day a Navy Officer called and asked me if I knew of a Navy lieutenant whose name I have forgotten. I told him that he and his wife were neighbors when we lived in Mtn. View, CA. He asked if he could come to my apartment and ask me a few questions, and I told him that would be O.K. I was busy repacking my wooden box for shipment when he arrived. The apartment was a mess, and I had on some old clothes that I had not bothered to iron when he arrived. I quickly cleared off some chairs to sit on. He was polite and asked a few questions about the former neighbor, but I really did not know much about him. After he left, I remembered that Stretch was in a squadron that probably required a top security clearance. I have often wondered if he was sent to check more on me than on a former neighbor that I hardly knew.

I had promised a friend that I would take care of their three children while they went on a three-week trip to Europe. I hated to call and tell her that I had received my letter. Claire understood that my place was with my husband and said she would ask another friend.

I had the wooden box ready for pick up in time for it to be shipped on the USNS *Barrett*. I packed my suitcase, and a friend drove me to San Francisco where I checked in at the Federal Building. This was the real beginning of my adventure. I soon found out that I would share it with a Navy family I already knew. Mary Jane Edixon, whose husband Jim was also a pilot in Stretch's new squadron, VW-3 Det.A, was there with baby Joanne. We had lived on the same street in Pensacola, FL, and played bridge often. It was wonderful to see familiar faces and to share the adventure.

I had received a letter from Stretch telling me that a Ruby Abbott would also be on the same ship. Her husband, Bud was another pilot in Stretch's Squadron, but was in the Navy Hospital Dispensary, NAS Sangley Point with possible hepatitis. It later turned out to be mononucleosis. His letter said that Ruby had two boys and a baby girl. Stretch asked me to find her and ask if she had received a letter from Bud before leaving Chicago. Mary Jane did not know Ruby, but we searched the crowd of dependents going overseas that had gathered in front of the Federal Building and found nobody to match the description of Ruby and family. Then a cab pulled up to the curb, and there they were! After introductions, I asked Ruby if she had received news from Bud about his problem. Stretch's letter had said that he was not sure whether Bud was to be sent back to the States, but that Ruby should check at the Federal Building for a possible letter from Bud. I offered to take baby Judy while Ruby and the boys went to see whether a letter was there. I held out my hands and Judy willingly came to me. I had no children of my own and was constantly borrowing those of everyone else.

The letter was there, and now Ruby had the problem of whether to proceed to the Philippines or return to Chicago. How was she to know if Bud was still in the Philippines or on his way back to the States? Mary Jane, Ruby, and I talked it over and concluded that she probably would not be allowed to proceed if her husband was being returned to the States.

Communication between our husbands and us wives was by letter and that could take a week or more getting to us. Cell phones were not available in the 1950's. Ruby decided that she would go to the Philippines as planned. All the paperwork, passport, and shots were completed, and unless told she could not go to the Philippines, she planned to make the trip. The boys, Bruce and Brian, were very happy that they would be going to see their dad. Baby Judy had never seen her father.

USNS *Barrett*, 1956

After receiving a briefing in the Federal Building, we boarded buses for the trip to the ship that would take us to the Philippines.

When we boarded the USNS *Barrett*, they gave us our cabin assignments and directions on how to find them. I found my cabin and that was a surprise: bunk beds were on opposite walls with a crib in between. It was cozy! A mother with two young boys was already in the cabin. She and the older boy had taken the bunk beds on one side of the cabin. The baby was in the crib. Ruthy Biggers, the other person assigned to our cabin, and I flipped a coin to see who got the lower bunk on the other side of the cabin. I won! Then we checked the bathroom and discovered that we shared it with four women in the adjoining cabin. That made seven women and two small boys using one bathroom! The mother and two boys were getting off the ship in Hawaii, so we agreed that we could manage in our crowded environment during those seven days.

My foot had been bothering me, and I decided to have the doctor on board the ship look at it before we sailed. He diagnosed it as a sliver imbedded underneath an infected callous on the side of my heel and proceeded to carve out the callous and remove the sliver. The doctor gave me instructions to follow during the healing process. I was to soak my foot in a bucket of salt water and bandage it only when going for meals or while sleeping. Sailing time was midnight. However, we were not to sail that night. The ship was experiencing mechanical trouble that delayed our departure. The next news informed us that we would be transferring to another ship. We lived on the USNS Barrett for four days before boarding the USNS General M. M. Patrick. During our stay aboard the USNS *Barrett*, I was stuck on the ship with my foot in the bucket of salt water while other passengers went ashore to tour San Francisco. Some of the wives stayed with me and they stacked suitcases between the bunk beds in the cabin adjoining ours and found two chairs. Somebody had cards and we spent time playing card games. Tim Kennedy, one of the wives in our adjoining cabin, was a nurse and she kindly tended to the bandaging of my foot when it was necessary for me to walk to meals.

Four days later the USNS *M.M. Patrick* docked beside the USNS *Barrett*. When it came time to move to the other ship, I had no shoes or slippers that would fit over my bandaged foot and going barefoot was out of the question. Mary Jane had a pair of scuffs and I borrowed one to hobble to the USNS *M.M. Patrick*. We nicknamed her the *Mickey Mouse Patrick*, and she lived up to the name over and over again as we endured our trip. We sailed at midnight, and I hobbled out on deck to watch the ship sail under the Golden Gate Bridge.

The following day I came down with the flu and did not recover until arriving in Hawaii. Someone had come aboard with the flu, and it spread throughout the ship. A child came aboard with chicken pox, and that spread among the children on the ship.

I was so sick that every time I had to use the bathroom I barely made it back to my bunk before I passed out. The mother of the two boys was also a nurse, and after several days of my not being able to retain water, she offered to help me to sick-call the next morning. We took burp bags with us, and I used them frequently. I made it to the sick-call waiting room before I passed out. I received immediate care of a shot and some kind of medicine and was back on my feet in time to enjoy our stop in Hawaii.

My laundry was piling up, and it was time to find the laundry room on the ship. The washing machines worked fairly well, but the dryers were a problem. For some unknown reason they worked best if someone sat on the lid. If no one sat on the dryer, the clothes were still damp when the machine stopped. We could take the damp clothes to the ironing room and iron them dry, but that was more time consuming than sitting on the dryer. Once while sitting on the dryer I met three Seabees, also doing laundry. They were also heading for the Philippines, but to Naval Station Subic Bay rather than to ours.

Two Navy nurses were on the ship heading for Tripler Military Hospital in Hawaii for duty. After my visit to sick-call, they checked on me frequently and we became friends. While we were docked in Hawaii, they cordially invited me to join them for a tour of the island as they had one more day before reporting for duty. We rented a car and drove around the island. That was to be the only time I saw any of Hawaii since my future trips to and from Japan and the Philippines were by plane and I did not get any further than the airport. What fun we had taking in the beauty and activity of that island! Moreover, it was a pleasant reprieve from the confinement of the *Mickey Mouse Patrick*! We even got to take a relaxing swim at Waikiki Beach and were able to get a tourist "gawk" at celebrity movie actor Gabby Hayes at the Royal Hawaiian Hotel.

Back at the ship, it was time to say goodbye to my two friends and thank them for a wonderful day. As I walked down the pier and up the stairs to the *M.M. Patrick*, I turned to wave farewell to the nurses. Then I threw my lei into the water as a goodwill sign that I would return to the Islands.

The cabins assigned to the dependents were located on the deck above the water line and troops billeted below decks. We also had separate fenced areas on the upper decks where we could enjoy the sunshine as we sailed south and west. Departing from Hawaii, we headed for Guam. A bridge tournament had been going on during the trip from the States to Hawaii. It was necessary to reorganize this because many of the players had left the ship in Hawaii. The Captain and I became partners. I had just learned how to play bridge when living in Pensacola, and told the Captain that I was a beginner. He was a jovial man and helped me with my game. We always played in his cabin so I did not have many other players available to ask for help, other than the Captain and the players in our foursome.

One day when I went to the laundry room, the three Seabees were there and they were good-naturedly teasing one of the group. The person on the receiving end of the heckling had bought a red flowered Hawaiian shirt and washed it with his other clothing. His underwear was a lovely shade of pink!

Food on the ship was terrible. Nine of us wives who did not have children sat at the same table in the eating area and we decided to try to do something about the food. Our plan was to have the cook come to our table and eat with us, and if he could eat the food, we would stop complaining. The evening meal when he was able to join us was Chicken and Dumplings. The chicken was so tough we could hardly chew it and the dumplings would bounce off the floor. The food did improve slightly after that, so we stopped complaining.

The tropical weather was enjoyable, but now we had no wind blowing over the ship. The sea breeze was blowing at the same speed as we were sailing, but the daily practice of blowing the stacks resulted in soot accumulating on the decks. The crew no longer hosed them off. We inquired about why the decks were not being kept clean, and it was explained that the soot was greasy and would make the decks slippery if they tried to wash them. Oh well!

An albatross landed on the fantail of the ship and was unable to take off. The big bird needed a good breeze to get airborne, and with a following wind, we had only a faint breeze. We threw chunks of bread to the albatross and he devoured them. He stayed on the ship for several days and then one morning was gone.

After several days of sailing the ship arrived at Guam, and I spent the day at a beach with some of the other passengers. We brought a picnic lunch with us and had a nice time swimming in the beautiful warm water and lying on the white sandy beach. The palm trees swayed in the breeze and the perfume from the tropical flowers wafted over us. I looked forward to living in the Philippines and wished we could hasten on our way. I longed to share this experience with my husband.

One day while I was sitting on the dryer, I heard a plane circling over the ship.

The dryer finished and when I got up on deck, the plane had disappeared. Mary Jane and Ruby thought it might be our husbands flying to meet the ship. Later I learned that it was true, but our husbands did not think we would be on such a decrepit ship. It looked like an old rusty freighter.

We approached the Philippine Islands from the East, and it was necessary to sail through a passage between islands. The scenery on the lush mountains of the islands was beautiful. The entrance to Manila Bay is on the western side of the Philippines. As we entered the Bay, I saw the island of Corregidor, and off to the left the mountains of Bataan. I remember reading about those places from news reports during World War II. A cousin of my father was an Army nurse on Gen. McArthur's staff and was able to escape from Corregidor with the General before the Japanese occupied it.

I could see the city of Manila in the distance, but what really shocked me were the numerous bombed out, rusted hulks of ships sitting in the mud of the Bay. Those rusting ships were now home to families of fishermen who had erected makeshift shelters on the decks. The water was muddy and debris floated everywhere. No beautiful blue tropical ocean greeted me. As the ship neared the dock, wives gathered on deck searching the crowd on the dock for our husbands. Shouts of greeting echoed from ship to shore, and we waved our arms as we each located our spouses. Then I noticed the buildings of the city. Many remained destroyed and some were only half rebuilt. I shuddered at the sight in spite of the warm tropical air. My joy in rejoining my husband took away some of the shock I witnessed of the destruction left from the war.

I had made a special outfit to wear to show off for my husband, but when I tried it on it hung like a sack. I had lost nineteen pounds due not only to the flu but also to the lousy food on the *Mickey Mouse Patrick.* We had been at sea for 28 days. I left the baggy dress in my suitcase and wore something else that fit a little better. I don't think he noticed what I was wearing. We were so overjoyed at being together again.

Stretch had made reservations at a hotel in Manila for my first night in the Philippines instead of taking me directly to Cavite City, which is adjacent to the American Base, NAS Sangley Point.

I had gradually become accustomed to the tropical weather while on the ship. The hotel room did not have air-conditioning and, unbelievably, I missed the hum of the ship. Stretch and I talked far into the night before we finally fell asleep. I tasted mango for the first time at breakfast the next morning and fell in love with the delicious fruit.

Now an insight into what our life in this new assignment would be. On the hour's drive from Manila to Cavite, I saw many nipa huts. They were the grass shacks with thatched roofs that the natives live in. I saw carabao lying in the water ditches beside the roads. Maybe that is why they call them water buffalo in some countries. Stretch tried to convince me that we would be living in a nipa hut, but I found that hard to believe. In areas where ground was prone to flooding, the nipa huts were built up on stilts. Others were at ground level. Very few had windows or screening. They were open to the elements. Chickens, pigs, goats, and carabao lived in the yard either under or around the nipa huts.

There was so much more to see on the road to Cavite City. Jeepneys were the primary means of transportation for the people of the Philippines. These are colorfully painted Jeeps left over from the war. Jeepney shops convert the cargo compartments of the Jeeps into seating for passengers. People and animals use this method of transportation the same as we use buses in the United States. I saw pigs, chickens, and even a small carabao in these Jeepneys. As we approached the city gates, we drove by some salt flats. At high tide, sea-water was allowed to flood the salt flats. Then the gates closed so no more water could enter the salt flats. The water would evaporate in the hot sun and the salt remained. The salt was raked up and sold for use by the natives. The bright white of the salt was startling in the tropical sun.

No guard was at the Guardhouse at the entrance to the city during the day. Stretch explained that there was a guard on duty during the night. Many people rode bicycles, others walked on the dusty roads, and a few rode in the Jeepneys. Just after passing the city market, we made a left turn in front of an establishment whose sign read, "Aroma Hotel." I thought, "How appropriate!" The city did have an aroma all to itself. I could smell garlic and onions cooking, and the brackish water of Bacoor Bay, an offshoot of Manila Bay, plus animal and human waste. People did not have bathrooms in their Nipa huts. They used their yards as toilets.

Picture is of a small hotel in Cavite City. We always thought the name was appropriate.

About two streets before the gate to the Base, we turned left onto a road called Pig Alley. Stretch stopped in front of a small compound and got out of the car to open the gates. "Here is our new home, no Nipa hut this time", he said. Dogs came barking, but quieted when Stretch spoke to them. They trotted off to lie in the shade of the caretaker's house. Frances, our "house girl," the Navy's phrase for a live-in maid, was there to greet me. Stretch had hired her to work for us. She was nineteen and I was just a few years older. It was her first job as a house girl and it was my first time to have hired help. We learned a lot together during my two years in the Philippines. Our house was the middle one of three houses on the right side of the walled compound. A bright green wrought iron gate was at the front of the compound. The entire compound was concrete, not a blade of grass anywhere. The caretaker's house was on the left. Our houses were two-story. A small living room, dining area, kitchen, and bathroom comprised the first floor. The stairway to the second floor began in the living room and went up to an open air porch where Frances did the ironing. Upstairs were two bedrooms. Frances lived with us and had her own room. Her only time off was on Sunday from 8:00 a.m. to 8:00 p.m. The bathroom was a square room that was completely tiled – floor, walls, and ceiling. The showerhead stuck out of the wall on one side of the room with no shower curtain. The sink and toilet were across the room. Stretch had rented rattan furniture for the house, since our furniture remained in storage.

One long stairway of 14 steps led to the second floor. Frances kept the floors and stairway polished with a dried coconut husk that was cut in half. She did a little hip-hop dance as she moved the husk back and forth with one foot and hopped along on the other foot. Polishing the stairs was done by using her hands to move the half coconut back and forth.

One day, as I was coming down the stairs, I made it to the second step from the top before losing my balance and bouncing all the way down the other twelve steps on my derriere. I sported a badly bruised backside and at the time had a white bathing suit. The bruises showed through the white material and my friends would accuse me of having a dirty suit.

The caretaker was also a fisherman, and we would often find the head and skeletal remains of fish on our front step. He would give the dogs some of the raw fish and they were very adept at eating the flesh without ingesting the bones.

Another Navy couple, Pete and Lee Chapola, lived next door and they already had their car. The Navy shipped our car by freighter from the Oakland Navy Yard to Manila. It did not arrive until several weeks after I did. Pete and Stretch were both in Squadron VW-3 Det A and Stretch would ride to work with him.

A night club in Cavite called Sheran Cove where we would go to continue partying after the O'Club on the base closed.

A few days after my arrival, a couple of other Squadron wives pulled into our compound to ask me if I would like to see the Cavite Market. After having a cup of coffee, they asked me if I was ready to go. What did they mean, 'Was I ready to see the

The Market in Cavite City

market?' I asked them to explain. They did not want to tell me too much about the place as they wanted me to make my own judgment. We drove the few blocks to it and found a place to park the car. The market was not much to look at from the street, but going inside turned out to be an experience I would not forget. I had to be careful where I stepped on the dirt floor so I would not trip over a ditch. Small stones lined ditches that channeled rain and other water from the street thru the market to the Bay at the back of the place. The roofing was such that it kept the rain off the merchandise but not off the shoppers. Stalls constructed of cast-off wood lined the walkways thru the market. The pathways did not follow straight lines and went off at different angles. There was no organized plan. Clothing could be next to dishes, and shoes found next to jewelry. The only signs of organization were the meat and fish area at the rear and fruit and vegetables near the street at the front of the market. Merchandise hung from the walls of the stalls or placed on old tables and benches. Many stalls sold ready-made clothes, jewelry, shoes, bolts of material, and hats. Others offered pots and pans, dishes, eating utensils, and all sorts of other cooking equipment. The market had no doors. It was a hastily built collection of stalls with corrugated tin roofs. I would consider it well ventilated. It was not raining on my first visit to the Central Market. I saw no evidence of electricity. The meat market at the back of the store was a real shock. Meat carcasses hung on hooks with flies and bugs swarming about. When a customer placed an order, the meat cutter gave the carcass a healthy swat and then began carving what the customer had requested. Then he placed it on an old-fashioned scale; the sale was completed and the money collected. The butcher wrapped the meat in newspaper and handed it to the customer. The flies and bugs returned to feast on the meat that was hung back on its hook. A large variety of fish was on display, but I did not see many people purchasing fish. I think they caught their own fish.

At the front of the market, next to the street the native fruits and vegetables were displayed in baskets. Bananas were ten centavos each. That was a nickel in American money. I did not make any purchases during my first visit to the Market. However, during my two-year stay, I returned many times to purchase fabric for dresses. I sent Frances to the market to purchase the delicious native fruit. It had to be soaked in a clorox-water solution before we could eat it. Most of our food was purchased at the Commissary located on the Base. We never bought any of the meat or fish from the Central Market.

The three seasons in the Philippines are the warm dry season, the cool dry season, and the rainy season. NAS Sangley Point was located on a spit of land that juts out into Manila Bay. This spit of land is about a mile long and that was mostly the runway. The base school was in a group of Quonset Huts, and the church was a large Quonset hut, across the street from the school. The hospital was in the same area, but was a cement building up on pylons. The Commissary, Navy Exchange, and snack bar were a block away. The only air-conditioned place on the Base was the photo-lab. Since Stretch had recently completed a course in photography, we spent a lot of time developing and printing pictures. Of course everyone stationed at the base wanted to move into base housing, but you had to wait till your name moved up on a list and that could take months, even a year or more.

The following is a story told to me by my husband regarding his top-secret new squadron.

Stretch really enjoyed flying these big black planes. The P4M-1Q planes the squadron flew were painted dark blue, almost black with no identifying marks. He could not tell me what the Squadron was doing on their patrols. Our husbands could not tell us when they would be going on a patrol, or when they would return. If the subject came up in casual conversation, the standard answer was, "You don't know, do you!" Then we knew that we had asked the wrong question. Sometimes three planes from the squadron would practice night landings and take-offs. Then one of them would leave and only two would land. However, Stretch did tell me that if a crew ran into mechanical or weather trouble and had to bail out or were jumped by MIGS, (Communist Chinese fighter planes), our government would not acknowledge their mission if they failed to come home. They carried no personal identification other than dog tags. The crews on these planes consisted of sixteen men, and they kept clothes and toiletries at various Bases where they landed to get some sleep, food and fuel for the planes. Such secrecy hadn't been a part of Stretch's Navy career thus far. I knew I wouldn't ask any more questions, but I had lots more in my mind.

We had been living in the house on Pig Alley for a couple of months and Stretch had been away on a long patrol. Manila glowed in the early morning and across the Bay, the runway lights of NAS Sangley Point were a welcome sight after a seven-day patrol. The landing was smooth in the pre-dawn warm tropical weather. After parking the plane and completing the necessary paper work, Stretch changed into his uniform and caught a Jeepney at the Base gate to get a ride home. It was just beginning to get light as he arrived at our compound gate. When opening the door to the house, he said, it sounded hollow. He shouted for me. Then he called to Frances. No one answered. He had a sinking feeling of being very vulnerable. He then remembered our conversation about moving to another compound

Pauly Varner in front of the Lodge near Lake Taal. It was a beautiful resort a short drive from the base at Sangley Point. Lake Taal is a body of water that has an island within the lake, and a smaller lake within and so on. If you counted them, there are seven lakes with seven island contained within.

called Certeza Court when a house became available. We wanted to get away from the house on Pig Alley, which was next to the Cock Fighting arena. The noise on Sunday afternoon was deafening as hundreds of Filipino men placed bets and watched cockfights.

In Pig Alley that morning, friend Pete Chapola heard Stretch calling for Frances and me and came to tell him that we had moved to Certeza Court. When I decided to rent the house there, Frances had hired six Filipino boys and a truck to move us to Certeza. Many of the houses Americans lived in were in compounds, considered more secure than single houses among the natives. Yes, our Pig Alley house was in a compound only a couple of blocks from the gate to the Base. Nevertheless, it was a very noisy place because of the cockfighting arena. On the other hand, Certeza Court was a much quieter location, near the Check Point into Cavite City.

Pete offered to drive Stretch to our new home. As they pulled into our driveway, Stretch was pleased to see our car. (It had arrived from the States while he'd been gone, and I'd picked it up.) Pete and Stretch came in, and after a quick look around, Stretch realized that he was very hungry, so I cooked breakfast for the three of us, then Pete headed back to the Base. Stretch then headed for a cool shower and to bed. There was no such thing as hot running water in the off-Base housing but a cool shower in the tropics is rather enjoyable. Our house shared a water tank with the one next door and sometimes on Saturday night Loplop, the Filipino caretaker of Certeza Court, would forget to turn on the pump to fill the water tanks. If we ran out of water flowing from the shower nozzle, we would turn on a faucet that was about a foot above the floor and squat down to finish showering.

Certeza Court was a walled compound of seven houses located on Bacoor Bay. This is a backwater of Manila Bay and the water is muddy, dark, and murky. Admiral Dewey was said to have hidden his ships in Bacoor Bay, but it is quite shallow and the large ships would have been mired in the mud. I doubt that the story is true.

The compound was fenced off from the water with barbed wire fencing. In the wee hours of the morning before dawn, we could hear fishermen drumming with sticks on the outside of the bonca boats. These are dugout canoes with a single outrigger. The steady rhythmic beating of the sticks drives the fish into the net, which is suspended from two long bamboo poles at the front of the bonca and lowered into the water.

A ten foot high stone wall with jagged glass imbedded into the top, and three strands of barbed wire above that, surrounded the compound on three sides. The fourth side was Bacoor Bay and that was fenced off with barbed wire. The front gate was open during the daylight hours, but the back gate was closed and locked most of the time. I was told that the Japanese used Certeza Court as a headquarters during World War II. An elevated watchtower was located beside the main gate. The watchtower was used by Loplop or his wife Mary when guarding the compound during the night. This discouraged burglars, who were often plotting to rob "rich" Americans. I did have my box of jewelry stolen during an afternoon when someone had left the back gate open and I was not home.

Three big houses that were closest to the bay had living quarters on the second floor. Rooms for the house girls and boys, space to park the car, and a storage area were on ground level. The three big houses had been there before the war.

At the back of the compound were four single story houses with carports and rooms for house girls at the ends of the carports. We lived in one of the smaller houses. In the far corner of the compound was Loplop's house where he and his wife Mary and their son lived. They had goats and chickens in their yard.

Entering our front door, one could see through the living room, dining room and kitchen to the back door. Our house also had two bedrooms with a bathroom in between. The ceilings must have been fifteen feet high. When I moved in, the hanging light bulb over the dining room table had burnt out. I would have had to place a chair on top of the table and then climb up to be able to change the light bulb. Bill and Mildred Deane lived next door. He was also in VW3-Det-A. Bill Deane was about 6'5" and he could reach the light bulb easily by standing on the table. Someone found a new bulb and he changed the light for me, since Stretch was gone.

At night we could hear machines guns going off up in the hills as the 'Huks' and Filipinos were fighting. The Huks were the communists in the Philippines and they were fighting the Filipino government, trying to take over running the country. The full name of that organization is Hukbalahap. At the time, Magsaysay was President of the Philippines, and he kept the Huks from gaining any power. We were safe within the city limits, but traveling could be dangerous, especially at night. We traveled to and from Manila by boat rather than drive.

Loplop's wife Mary patrolled the compound at night. We could hear her crunch, crunch footsteps on the gravel as she mumbled to herself. The Japanese had tortured her and it had affected her mind. I heard she carried a pearl-handled 45 to protect either herself or us, or maybe both. I never saw the gun, and Mary was unable to speak. She would give us her toothless smile, but that was all. We Americans accepted her as she was, but we could not talk with her.

Soon after moving to Certeza Court, Martin came to work for us as a houseboy. He was from the southern islands and spoke only an Ilocano dialect. Frances was from northern Luzon and spoke Vizayan. She also spoke Tagalog, which is the universal language of the Philippines. Martin did not understand Tagalog and would go to his cousin, who worked in the compound, to find out what Frances wanted him to do. She gave him the heavy work of polishing the floors, keeping the spider webs out of the rattan furniture, polishing shoes, washing and polishing the car, and keeping the yard neat and tidy. He was a very intelligent boy and was speaking English within a month of his arrival. I do not know where he slept at night but I assume he had a place with the other boys in the compound.

Certeza Court had beautiful large shade trees. I think they were mango trees but I never saw any of the fruit. The big tree that was near our carport had a huge spider web in it. I think it was about six feet in diameter. It swayed back and forth in the breeze with a very large spider in the center. I never saw the spider leave the web and I was happy that he remained in the tree. Geckos (small lizards) and bugs shared our house. I had to chase the geckos out of the shower stall when I bathed. Because of the tropical heat, the bed only had a bottom sheet. I would stand in the center of the bed using a can of bug spray and make a 360-degree turn to disinfect it before lying down for the night. If I forgot this routine, I would have bites from the tiny spiders that gathered on the bed.

Finally, after months, we moved up the housing list far enough to qualify for Base housing, everybody's dream. Martin and Frances still worked for us when

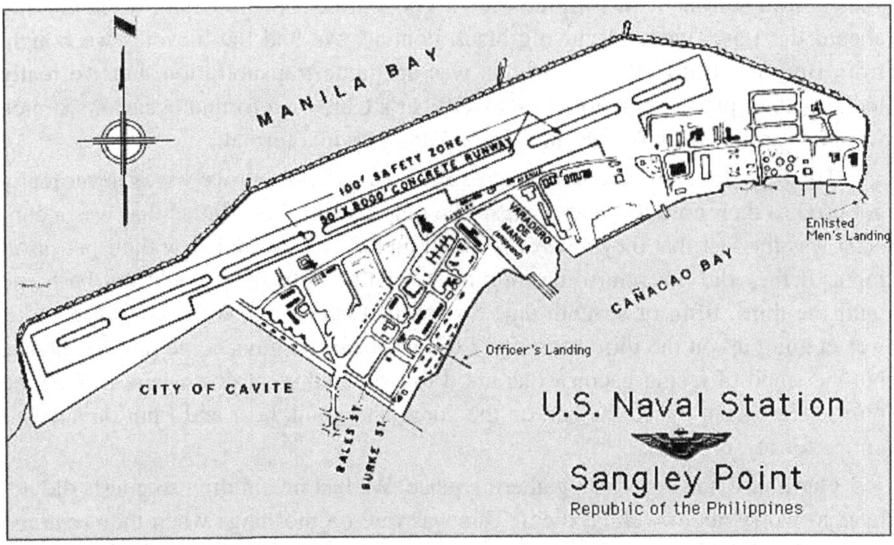

we moved from our one story house in Certeza Court to Base housing. I explained to Martin that when we moved into the two story house on Base I did not want him to polish the stairs. That stairway had a landing halfway down to make the turn into the living room.

One day I was in a hurry and came home to get something from the upstairs bedroom. Being young and athletic, I took the stairs two at a time. Coming back down I slipped on the landing and bounced the rest of the way into the living room. Martin was there polishing the floor. I was a sight with my skirt up around my waist.

I looked up at him and said, "Martin, you waxed the stairs?"

"Yes, mum," he replied.

I gave him a stern look and reminded him, "You don't wax the stairs."

"No, mum!" he said as he shook his head.

Then he quickly came to help me up and with concern in his voice he asked, "Are you hurt?"

We had moved into the house on Base that Bud and Ruby Abbott had occupied until they returned to the States. We also bought their bicycles. The first floor had a nice large kitchen, living room and dining area. A large screened in porch was at the back of the house and there was a big storage room in the carport where we put our personal refrigerator. The Navy furnished one in the kitchen. We kept beer and soda pop in the refrigerator in the storage room. The second floor had two bedrooms and a large bathroom with running hot water. Frances now lived in a dormitory on Base and not in the house with us. I do not know where Martin lived, but he only worked for us three days a week and another family three days a week.

The law was if you had a car shipped to the Philippines, it had to be there for a year before selling it to Filipinos for a good price. Not too long after moving aboard the Base, we sold our big black Pontiac. We had the bicycles we bought from Bud and Ruby Abbott and this was adequate transportation, but we really looked out of place when going to the Officer's Club on a formal occasion. Stretch would wear his Dress Whites and I would dress up in a formal.

The Squadron continued their top-secret spy patrols, but we wives never really got used to their comings and goings. The only part of their flights that was a constant was the fact that they always returned on an odd day and flew their patrols at night. If they did not return the morning after they left, they would not be home until the third, fifth, or seventh day. Stretch had been gone for a few days and it was coming up on the third morning. I did not hear the guys come into the house, but the smell of frying bacon woke me. I dressed and went downstairs to feed the pilots. They went out to the bar, on the porch and drank beer and I put on a pot of coffee for myself.

Our house had become a gathering place. We had no children so guests did not have to worry about waking them. This was true on mornings when they returned from a patrol as well as after the bar had closed at the Officers' Club.

One of the pilots lived off Base, and this particular morning Stretch offered to give him a ride home. This puzzled me because we only had our bicycles.

"Come with me to the carport," Stretch said as he led me through the kitchen. Lo and behold, there sat a vintage 1930's Chevrolet. I had not heard Stretch drive the car into the carport due to my hearing problem, but there it sat. He had bought it for $200.00. It had no window on the driver's side and the shift lever was on the floor between the driver and passenger. It was quite the car, and I quickly became accustomed to its quirks. Before getting in, I would open all the doors and bang on the seats to drive out the geckoes. The first time fellow Squadron wife Alice Sumner saw the car, she named it, "The Brown Pile".

I drove it around the Base but did not really trust the old car enough to take it into Cavite. It was a bit tricky to get it moving in first gear; it did much better in second or third. The road out to the Main Gate had a stop sign that required everyone to stop before proceeding out the gate or making the right turn to the swimming pool where I was teaching regular swimming classes. I was used to slowing down and coasting around that corner in second. That is, until I was caught. Yes, I received a ticket and was told to show up at the Marine building to discuss my actions. I think "Locker Box Jonesy," a friendly Marine Officer, had set a trap for me. He would often stop by the pool during the Boy Scout Life Saving classes to see how the sons of dependents were doing. The day of my hearing, he was not in his office. I was explaining to the Marine behind the desk why I thought the stop sign should be on the other corner when I heard someone laughing outside the door. I boldly spoke up and said, "Jonesy, are you in back of this ticket? Entering the room, he put his arm around my shoulder and gave me a hug. Then he told a young Marine, "Go, change that stop sign."

Before we left the Philippines, I sold the car to another couple for $200.00, but the title of "The Brown Pile" stuck with it. It was just an old "beater," but its four wheels and enclosed body sure made grocery and Base Exchange shopping easier than our two-wheeled bikes. Its roof kept out most of the tropical rain.

We bought a rattan bar and four bar stools for the porch and added some lounge chairs to the area. This became a gathering place after the Officers' Club closed for the night. We lived a rather upside down life. Partying lasted way into the night and often ended up with a breakfast of bacon and eggs before people left for home. During a party when Stretch was perched on his bar stool playing bartender, he would ask me to bring more beer, soda, or ice from the outside refrigerator. I would pick up the carrying basket, catch the eye of one of the wives, and give the signal, 'I am heading for the carport'. This would lead her to join me for protection. One of the men seemed to know when the bar needed supplies and would be out in the carport to try to hustle someone else's wife. We were wise to this Romeo, so we made the trip in pairs. This prevented him from making any captures.

John and Helen Ebersbaker decided to host a party at their house on the Base.

John was one of the COM (Communications) Officer's in the Squadron. John and Helen thought it would be a good idea to mix the Martinis and Manhattans ahead of time. They mixed up large containers of these drinks, and put them in the freezer. However, they neglected to add any ice to these delicious drinks. The guests at the party were unaware of the potency of the alcohol in these drinks. Soon the party was in full swing. Irma Kruzich had one of those Suzi Wong dresses and a group of the men decided that the slits on the sides of her dress were not high enough. They found a scissors in a kitchen draw while a few of the other men escorted Irma to the stairs that led to the upstairs and sat her down to do their cutting. She was a good sport and didn't put up much resistance. The men proceeded to cut, as carefully they could, the stitching and not the material. I don't remember just how high they went with their cutting. The party ended early. Stretch had the duty that day, and that meant no drinking for him. I had been ill most of the day so I was not imbibing either. Stretch offered to drive Don and Alice Sumner home and I would follow in our car. Don decided for some unknown reason that he did not want to ride with Stretch and his wife and crawled into the back seat of my car where he promptly fell asleep. When we got to where Don and Alice lived Stretch pulled into the compound to let Alice off while I pulled up around the corner to wait for Stretch to come and take Don home. Alice was not very happy with Don, but then after a few weeks of the silent treatment they made up and returned to be the loving couple they were. Irma's dress was repaired and John apologized for his lack of knowledge as a bartender.

One afternoon I sat talking with Stretch about what I could do to make myself useful while living at NAS Sangley Point. Tea parties and bridge playing had become disenchanting. I had servants to do the housework. Letter writing and sewing filled some of the time but I wanted to do more. Shopping in Manila was fun, but there were a limited number of stores. We knew that I enjoyed swimming and together we came up with the idea of starting a swim program. The next day I made an appointment with Mr. Yale, the Red Cross Field Director at the Base. He had books on how to go about setting up such a program and presented me with an armload of these books. I was a qualified Red Cross water safety instructor and knew how to teach swimming, but I had a lot to learn about how to go about setting up a swim program. I had frequent meetings with Mr. Yale, and together we came up with a program for training people to help with the program. I began writing articles in the Base newspaper and set up a meeting for those interested. Several of the Squadron and Base wives came to our first meeting. Many of them had been water safety instructors during their college years. I sat down and wrote two class schedules, one for retraining and updating the women who had been qualified instructors. The other was a Swimmer Aide course for those who wanted to help but had no previous training. The weather was great for almost year round swimming and an Olympic size pool was available for use. Now we were ready to start classes for the dependents.

My next task was to write newspaper articles to find out how many children and adults were interested in learning to swim. The Base school also sent home questionnaires to find out what kind of a program I needed to schedule. The huge response was overwhelming. I learned that the biggest need was for a Beginner program, so I devised a method whereby such a program could be broken down into small components. I divided the beginner skills into ten Stations. Station 1 was for those children who were learning to put their faces in the water, blowing bubbles, and learning to float on their stomachs. Station 2 was for those who were learning how to kick their feet and use their arms. Station 3 was for those learning to back float and propel themselves while on their backs. When they reached Stations 9 & 10, they were given the Beginner Tests and then progressed to an Advanced Beginner course. This enabled us to accommodate more children during the time we could use the pool. The instructors taught the more advanced classes and the aides were utilized in the Station-to-Station method for Beginners. I kept track of these beginners by having the aides tell me who in their group was ready to progress to the next station. I would go home and type up the class schedules for those children for the next day. It worked beautifully. The children were learning, and the instructors and aides were happy to be doing something productive. Adults were asking for classes for themselves, and I scheduled them, since I knew that we had to have new instructors and aides to replace the ones that were returning to the States. I now felt like I was contributing something useful for the dependents.

One of the Squadron wives, Jinny Martin, approached me with the story of how she had been shoved into a pool by an instructor. She admitted that she was terrified of the water. Could we teach her how to swim? With much patience and encouragement from all the instructors, we helped her to overcome her fear of the water. Within six months, she had passed all the requirements to take the instructor course. She became one of our best instructors and was particularly good with students who were afraid of the water.

The Boy Scouts asked for a special program for their requirements of a badge and I was able to schedule that. The Girl Scouts heard about it and wanted a similar program. I convinced them that a combined program would be better. The evaluation for the Boy Scouts and Girl Scouts is in their handbooks. Both water safety instructors and Scout leaders collaborate on passing the child on their skills. That was a fun program for me as well as the children. I had to listen to songs and poems that they had written about their school teachers before starting class each day. They learned how to swim with a backpack heavy with rocks for survival training in addition to making simple rescues and self- rescues.

Some fathers approached me with the idea of starting a swim team. I agreed, with the understanding that they teach me how to do this. I had no knowledge of swim team competition, and they were most helpful. Every afternoon just before dinner, the men and the swim team group met at the pool for practice. Soon we were ready for our first meet. One father offered to make and install lane lines in the pool; another said he would take on the task of getting medals. A couple of them agreed to be stroke and turn judges and several volunteered to be timers. The turn and stroke judges watched the competing swimmers for errors. Turns for each of the strokes used in competition must be performed correctly. No extra pull of the arms or extra kick is allowed during a breaststroke turn; this is something watched for carefully. No deviation from any of the strokes is allowed. Where the stopwatches came from I never learned, but at least fifteen were put to good use. We needed a minimum of fifteen. Three timers on each of the five lanes. These men now went to work timing and organizing the team into freestyle, backstroke, breaststroke, butterfly, and relay teams. Both boys and girls joined the swim team.

Girl Scouts asked for lessons in water ballet. I had been on a water ballet team at the YWCA when I was a teen-ager, but knew nothing about how to teach these skills or organize a performance. I told them I would write to my mother to see whether she could find some books on the subject. A couple of weeks later two books arrived and I began teaching the girls these skills. They wanted to be ready to perform the day of the swim team competition. When the day came for these events, the girls and parents were busy gathering hibiscus, frangyipani and other flowers for their performance. Leis were constructed, bathing suits donned, flowers put in their hair, and the leis draped around their necks. They giggled with anticipation as the people arrived for the big day. Marching out onto the pool deck, these girls took their places at the edge of the pool. A peel-off dive was performed and when the girls surfaced, the pool was afloat with flowers. Just for a moment, they gave me a quizzical look but as soon as they heard the music, they went on with the performance. The audience gave them a standing ovation. The floating flowers were retrieved, the lane lines were installed and the swim team started their competition. The swim team agreed to have a meet every six months. I was in the process of finding out whether Clark Air Force Base and Subic Bay Naval Station, other nearby U.S. bases, had swim teams when Stretch received orders to the Naval Post Graduate School in Monterey, CA. This meant we left Sangley Point and I never heard whether these other Bases had swim teams.

Also during my Sangley stay, I had the opportunity to become a First Aid instructor. I eagerly joined the program. Mr. Yale called me one day to tell me that a Red Cross conference was being scheduled between the Philippine Red Cross and the American Red Cross. It would meet for a week at Clark AFB followed by three days at the Manila Yacht Club. The subject at the conference would be Water Safety, First Aid,

Pauly Varner paddling a Bonca boat with Philippine army man

and Boating. Would I like to attend the conference? I checked with Stretch, and he advised me to apply for a chance to participate. I was able to get transportation on the daily mail plane. It was an amphibious plane that departed from Sangley Point in the morning with a stop at Naval Station Subic Bay before arriving at Clark AFB. We took off from the runway at Sangley and landed on the water in Subic Bay. A boat from Naval Station Subic met the plane and mailbags were thrown into the boat. Passengers then climbed down into the boat. After the boat routines were finished, we landed on the runway at Clark.

The attendees at the Red Cross Conference were six Air Force wives, six Filipino Army men, Mr. Charles Russell, and me. They gave me the nickname of Miss Sangley Point, as I was the only Navy wife. Mr. Russell, the leader of the conference, flew in from Tokyo where he was the American Red Cross Field Director in charge of Safety programs for the military in the Far East. At the conclusion of the conference, Stretch and I invited the attendees to our home in Certeza Court for a barbeque. The group traveled from Manila to Sangley by boat and then by car to Certeza Court. One of the Filipino men was called Cinco, not his real name. He was born on May 5, married on May 5; he had five children and was five feet tall. A buddy of Cinco told us that he knew how to perform magic. We all wanted to see his magic tricks. Cinco announced that he could put a whole banana into an empty bottle of scotch. The group had been imbibing and the empty bottle was located. Next was the banana (peeled of course), and a match. A few drops of scotch remained in the bottom of the bottle and he announced that was good, as he did not want to waste even a drop of the liquid by having to put a few drops back into the bottle. Placing the bottle on the picnic table, he lit the match with one hand while holding the banana in the other. Quick as a wink he threw the match into the empty bottle and even quicker placed the banana on the bottle. Yes, the whole banana was now inside the bottle. We all wanted to see it again, but first we had to empty another bottle. Hands holding empty glasses appeared and we soon had an empty bottle for the next magic performance. That whole week taught me a lot and I had lots of fun too.

When our husbands were able to take leave, we traveled to Hong Kong and Baguio. The approach and landing at the old Kai Tak airport in Hong Kong was thrilling as the pilot watched for the checkerboard painted on the face of the canyon wall. He then made a hard right turn and flew down the canyon to the airport. The wingtips of the plane almost touched the walls of the canyon.

When going to Hong Kong for a three-day R & R trip, we brought along pictures of clothes we wanted made, and our husbands ordered tailored uniforms, monogrammed shirts, and Barong Tagalogs. Barongs are wonderful dress shirts worn by men. They are made from Pina cloth, a fabric made by scraping the fibers from leaves of the pineapple plant with a broken piece of pottery. Then the fibers are tied together with invisible knots and woven into a light ecru colored fabric. A fine white cloth is sometimes used as the fabric of choice. The Barong is beautifully embroidered on the front alongside the buttons and on the cuffs. The Barong is worn loose outside the pants and is a much cooler type of formal clothing than military dress uniforms.

We also enjoyed many points of interest on the island of Hong Kong and the mainland city of Kowloon. The cable car trip to the top of Mount Victoria will forever be in my memory. The cars fit the slope of the mountain. The seats were level but the cars were on a slant, giving the passengers a perspective similar to being on bleachers at an athletic field. The view from the top was magnificent. We could see the entire Bay of Hong Kong, with ships, ferries, fishing boats, sampans, and houseboats. Sitting on the balcony outside our room at the hotel and sipping martinis, we watched the ships entering the harbor. The Hong Kong newspaper gave a listing of ships arriving each day, I used the small binoculars I carried to determine the country of origin of the ships. I had a hobby of learning the flags of countries around the world.

Another tourist place in Hong Kong was the Tiger Balm Gardens. This beautiful garden had been built with money donated by the man who manufactured and marketed the balm, a very aromatic healing ointment, made from the resin of a tropical evergreen tree. It is sold all around the world. The garden is very elaborate with both a Chinese theme and English architecture.

I remember being in Hong Kong the day the first American prisoners were released from Communist China. The Communist Chinese government did not allow foreigners into China and those who managed to enter China were put in prison. Negotiations between our government and Communist China made it possible for these nine or ten prisoners to be released. Only a few select people were allowed near the area where the prisoners walked across a bridge to Kowloon. At that time in history, Hong Kong and the mainland city of Kowloon were ruled by the English. We could hear the guns booming forth in celebration of the occasion. I later learned that it was Dick Tomlin, the American Red Cross representative, who greeted those prisoners. I had met Dick at Mr.Yale's Red Cross office in NAS Sangley when I was in charge of the swimming program there. A few years later, we became better acquainted with Dick and his wife Hazel while we were both stationed at Marine Corp Air Station (MCAS) Iwakuni in Japan. Between 1956 when we left NAS Iwakuni and our return to that base in 1962 it had been changed from a Navy base to a Marine Corp Air Station (MCAS). We lived in the town of Iwakuni for 14 months before moving to housing on the Base. Stretch was assigned to VP-50, a seaplane (P5M's) patrol squadron. Dick was the Red Cross Field Director there, and I was in charge of Safety Programs.

Another favorite place to R & R was Baguio, the summer capital of the Philippines. It is high in the mountains north of Manila, where it is cool and pine trees grow. Camp John Hay, near Baguio, was a recreation retreat for United States military. We would check out a Special Services station wagon from Sangley to drive to Baguio. Stopping at Clark AFB enroute to Baguio for lunch and shopping was a welcome respite before heading up the switchback mountain road to Camp John Hay. I later learned that the Filipinos have a bicycle race up that mountainous road that has about 36 switchback curves! On one of our trips to Baguio, a typhoon hit the area and we were all huddled in the dining hall until the storm was over. Another time in Baguio, when we were sharing the trip with another couple, we learned that a road led west down the mountain area to the Lingayan Gulf. We had learned about that place in high school when WWII was going on. We wanted to see this historical place because it had been one of the landing beaches when American forces returned to free the Philippine Islands from Japanese occupation. After getting a box lunch prepared for us at the camp dining room, we followed directions to the Lingayan shore. Here was the beautiful tropical beach with the palm trees swaying in the breeze and wild bougainvillea in full bloom. The only flaw in the idyllic scene was the rusted old landing craft in the water near the right side of the half-moon beach. Thatched roof dressing rooms were available for us to change into our bathing suits. We raced for the beautiful blue water only to discover it was infested with thousands of jellyfish. That ended our swim for the day, but we found a picnic table and unpacked our delicious lunch, posed for pictures, and relaxed on the beach.

There was much more to learn about Filipino history in and around Manila. We visited Malacanyan Palace where President Magsaysay and his family lived. We could stroll around the beautiful gardens surrounding the palace and some of the rooms inside which were open to visitors. I attended a wonderful luncheon there hosted by the Philippine Red Cross for members of the American Red Cross. The food was delicious, but I do not remember the menu as it was many years ago.

Another historical place of interest was the Santiago Prison, where chains hung from the walls. This is where prisoners were tortured by the Japanese during WWII. The walls that surrounded the old city of Manila were still there on the banks of the Pasig River. We could walk along the top of one wall. At the center of the city, from which measurements for all distances in the Philippines are measured, is the Cathedral, which was completely bombed out. Papaya trees grew up on the altar and natives had erected lean-to shelters along what was left of the walls. Children played among the ruins and music blared from a radio. Small cooking fires burned in the rubble in preparation for the next meal. Chickens, pigs, and goats roamed free in search of food. Ten years later, when we returned to the same base, we went to see the Cathedral. It had been beautifully restored, I am happy to report. We visited another church in Manila that is reported to be over 900 years old. It had big double wide, thick wooden doors that were beautifully carved. The doors were locked so we could not see the inside. That church has never been destroyed by wars, earthquakes, or fires.

We traveled to and from Manila via a crash boat. It was so called because if there were an accident with ships or a downed airplane near or on the water, that boat would be sent to help. This trip took about an hour unless American carriers were in port. The carriers could not pull up to a pier in Manila because the water in Manila Bay is shallow and many half-sunken ships littered the water. About 750 Japanese ships had been bombed and sunk in Manila Bay by the American Forces when they liberated the Philippine Islands. The carriers anchored out in Manila Bay. The boat from Sangley Point stopped at these carriers to pick up or drop off passengers on the way to and from Manila and Sangley Point.

We often received an invitation to have dinner on board a carrier, since our husbands knew some of the pilots stationed aboard. We had to wear dresses; no slacks or shorts were allowed beyond our quarters. Getting from the crash boat onto the ship required athletic ability. First, the men hoisted us up onto the water barge, then we walked across the barge to the stairs. The water barge is a large flat barge that is filled with fresh water and towed out to the carriers to re-supply the ships. The location of the water barge prevented extending the stairs all the way down to the surface of the water. The treads on the stairs were at a 45-degree angle, so we had to balance ourselves on the forward edge of the stairs. If we wore heels, the walk up those stairs was quite a challenge. We also resorted to wearing shorts under our dresses, as it could be quite breezy out in the Bay.

Another activity we enjoyed in Manila Bay was R & R trips aboard the *Margaret* to Corregidor. The *Margaret* was a 65-foot pleasure boat operated by the Special Services Dept. NAS Sangley Point for recreational use by the people of the Base. I do not recall the story of how it ended up in the Philippines, but it was wonderful for us to be able to use. Corregidor is the island where General McArthur made his last stand before escaping from the Philippines. Located at the mouth of Manila Bay, it has tunnels carved into the hills, and we were able to walk through these caves. Touring the island in old Army stake bed trucks was an adventure. Rusted artillery guns and live ammunition were everywhere and many of the guns pointed skyward as if shooting at planes. The deserted parade grounds at the once active Army base were rapidly being overtaken by tropical growth. The Mile Long Barracks where the U.S. Army was billeted before WWII was in shambles and not safe to enter.

NAS Sangley Point was the closest American Base to the American Embassy and therefore was often called upon to host receptions and parties for visiting dignitaries. At one of these parties honoring Admiral Radford, the food served was native to the Philippines. Food that we Americans had never heard of we tasted and enjoyed. I remember barbequed snails on tiny sticks, octopusi, and many types of fish whose names I don't remember. The next day some of the people from our Squadron went on an R & R trip to Corregidor aboard the *Margaret.* It started out as a beautiful day, and the Bay was sparkling blue. Then it hit! We called it the Cavite Bug, and almost everyone aboard came down with extreme nausea and diarrhea. The bathroom facilities on the *Margaret* were minimal and people were miserable. Soon the boat turned around and took us all back to NAS Sangley Point; then everyone headed for home. Only those from our Squadron and others from the Base who had attended the party suffered the malady. Stretch and I had not eaten much at the party and escaped the ravages of the aftermath.

We would often take a day trip to Lake Taal, which was always fun! The lake is in a volcano about an hour's drive from Manila. From the lawn in front of the

Lake Taal Lodge, one can see that there is an island within that lake. If you fly over Lake Taal, you can see a smaller lake on that island which in turn has an island in it. The story goes on, with a lake within an island for how many times I am not certain since I have never flown over Lake Taal.

During one of our trips to Lake Taal, I saw a performance of the Tinikling. This is a dance using two long bamboo poles and two short bamboo poles. The long poles are placed side by side on the floor. The short poles are placed about a foot from the ends of the long poles and then inserted under the long poles to keep them lifted off the floor. Two people place themselves at the ends of the long poles to hold the ends. When music begins, the pole holders begin practicing a rhythm of two taps on the small poles followed by sliding the poles together with a slight bang. To prevent their fingers from getting smashed, they keep the fingers on the outside of the poles. After a few practice tries to get the rhythm correct, one or two dancers position themselves near the center of the poles, but on the outside. The music begins and the pole operators begin their tap, tap, slide, bang to the rhythm of the music. The dancers hop in and out between the poles without getting their ankles caught. This is a very graceful dance which requires a great deal of skill and is beautiful to watch, especially when the pace of the pole rhythms gets faster and faster.

In June 1955, the Squadron designation changed from VW-3 Det A, and was commissioned VQ-1 A few weeks after the commissioning the Squadron's homeport was changed. We would move from NAS Sangley Point to NAS Iwakuni, Japan. Preparations to move from the Philippines to Japan progressed rapidly.

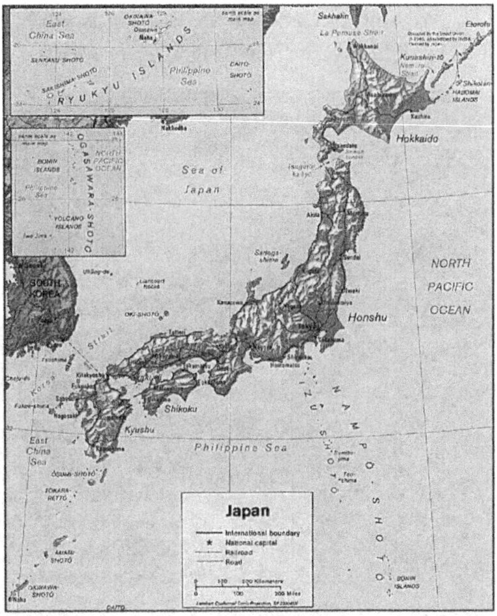

Just after the move was announced, Stretch and fellow Squadron pilots Don Sumner and Jim Edixon received orders to the Naval Post Graduate School in Monterey, California. They had requested an assignment to the Meteorology School and all three were accepted. Mary Jane Edixon was pregnant and was not able to join us in Japan, but Alice Sumner and I were able to go to Japan for the six weeks the men had remaining in the Squadron. The men flew the planes to Japan and the women and children were to follow. Those families living off Base were assigned to stay overnight with those living on Base at Sangley. Nancy Grau and LaVon North and son Gregg stayed with me. That evening word came that a typhoon had hit the area in Japan where we would be living. The sea wall had broken and the housing area damaged. That meant there would be a delay in our leaving the Philippines.

Since our gear was all packed, we borrowed pots, pans, and dishes from neighbors. Then we bought supplies from the commissary and began cooking those meals that we did not eat at the Officers' Club. Every morning we would get up, wash our clothes, pack our suitcases, wash the bedding and be ready to leave if we got word that the housing in Japan was ready. Our former house girl, Frances, already had another job, so we did our own work. This went on for a week before we finally received the happy news that we were able to go.

That morning our suitcases were picked up and loaded onto the waiting plane and our new adventure was about to begin. Jerry McMorrow, our military escort, joined us in the Snack Bar as we waited to board the plane. Several people whose husbands were attached to the Base came to say farewell. The announcement came over the loudspeaker to board the plane. That was followed by a second announcement telling us to return to the Snack Bar. Jerry went to find out the cause for the delay. It seemed that someone had forgotten to weigh our baggage. It was necessary to off load the baggage, weigh it and put it back aboard the plane. This is required to ensure that the plane has the proper balance for flight. The children sat on the floor and we read stories to them until it was time to get aboard the plane. One of the stories we read to the children was about Jerry McBoing Boing. Jerry a bachelor, listened to the story and then asked: "Why did I get the nickname, McBoing, Boing?" We assured him that it was not due to the story but just the guy's way of giving him a nickname.

We said our good-byes to friends we had made in the Philippines, and, after a weeks delay we were on our way to Japan.

NAS Iwakuni in 1956. Photo USN.

On the way to NAS Iwakuni, we landed in Okinawa to refuel. Or we were advised to stay there due to bad weather in Iwakuni. It was nice to be able to get out of the plane and walk around the air terminal, and get out some of the kinks from being on the plane for quite a while.

It was late afternoon before we landed at NAS Iwakuni and our husbands were there to greet us. Our Passports were collected, stamped and returned to us. That insured that we were now residence of the beautiful Islands of Japan. Some of the wives and children on the plane became ill due to having to wait in a holding pattern when we approached Iwakuni, Japan. I am not sure what the reason was for the delay in landing, but it caused a lot of grumbling. After dinner at the Club, we boarded enclosed boats to make the night trip across the Inland Sea to the housing area called Niji Mura. It was a long two hours before we docked in Kure and climbed aboard buses waiting to take us to our assigned apartments. Alice became ill on the boat trip, and Stretch and I agreed to take Nancy Sumner, age three, while Don helped Alice, baby Barbara, and Billy, age five. Nancy and her brother and sister called us Uncle Stretch and Aunt Pauly. Nancy was tucked in bed and sleeping when Don came to the door to get her. We assured him that she was fine and we would bring her home in the morning, but he said that Alice really wanted Nancy to be with them. We wrapped her in blankets and Don carried her to their new home. None of us had any cars as we had sold them in the Philippines.

It was October, and we soon learned that Japan's autumn climate was much cooler than that of the Philippines. Whereas we had been wearing sundresses and shorts at tropical Sangley, (even our husbands wore uniforms of khaki shorts to work), now in Japan, we needed cool weather clothing. Fortunately, Stretch and I had some warmer clothes stored at Sangley. Shortly before the move to Japan, we made a trip to Hong Kong and had some warmer clothes made.

The first day that Stretch did not have to work, we took the train to Hiroshima, an hour's ride. It was 1955 and Ground Zero, where we dropped the Atomic bomb, was as if that tragedy had just happened. The devastation was complete. Chunks of concrete and dirty dust covered the entire area. Not a blade of grass was visible. The only building standing was the skeletal remains of what had been the hospital. We took pictures and continued our walking tour. The Japanese are friendly people. They approached us asking if they could walk with us and practice learning to speak English. Rebuilding of the city was in progress, mostly with pick and shovel. We saw men working shoulder to shoulder with picks to rebuild the roads. They chanted a song that kept them swinging the picks up and down in a rhythm. The work was slow and tedious but they kept on working. We returned to Iwakuni in 1962 for a two-year tour. By then Hiroshima was a beautiful city and Ground Zero a beautiful park.

Shortly after visiting Hiroshima in 1955, Stretch was able to take ten days leave. We headed to Tokyo and also planned a three-day visit to Kyoto, one of the main culture centers of Japan. On the train from Iwakuni to Kyoto, we stopped in Osaka. After a stop at the train station, the train pulled out into the train yard and began unhitching the cars at the front and the rear of the train. The engine disappeared into the night with those cars. We sat there in our car, wondering if we were supposed to change trains to get to Kyoto. We spoke no Japanese, and we had just returned from having dinner in the dining car. After what seemed like an eternity, things started happening. Clanging and banging began and we became part of another train. We wondered if we were on the right train and heading for Kyoto. A small Japanese man in uniform came into the car to check our tickets. He bowed and with a big grin said, "Yes."

Our three days in Kyoto were wonderful. It was just like the stories I had read. We toured a big castle, climbed steep wooden steps in our stocking feet, and saw beautiful temples. At a Shinto shrine, we watched brightly costumed dancers perform. The factories where they made lacquerware and cloisonné vases and dishes were fascinating. I bought pearls from a delightful Papa-san who spoke English. He advised us on what to see while in Kyoto. We stayed at a beautiful Japanese inn. During the occupation, Americans could stay at these Inns for $1.00 a night. It was the first time we slept on a Futon on the floor. Futons are Japanese mattresses put on the floor to sleep on. During the day, they are stored in a closet. Both sexes use the same bathrooms. Rooms do not have individual bathrooms. That took a little getting used to but, "When in Japan, live like the Japanese." Before leaving Kyoto, we went back to the Mikimoto Pearl Store to thank the Papa-san for his help.

The train trip from Kyoto to Tokyo was very picturesque with rice paddies and fields of mulberry bushes for the silk industry. Then Mt. Fuji came into view. When I was in the fifth grade we studied Japan, and a classmate and I made a model of Mt. Fuji. I remember asking my mother for some sugar to use for snow on the top of the mountain. She said that salt would be better so it would not attract the ants as much. Now as an adult, I thought the mountain was beautiful and little did I realize at the time that I would return to Japan two years later. We lived on the base at NAS Atsugi and I could see Mt. Fuji from my bedroom window on clear days.

In Tokyo, we stayed at a military facility where we had a western-style hotel room with a bed and our own bathroom. We shopped on the Ginza, a district in Tokyo of small shops and street vendors, where Stretch had used cigarettes to barter for souvenirs during the Korean War. We ate delicious food and attended shows and dance performances. Our suitcases were bulging with purchases when we boarded the overnight train back to Iwakuni. Stretch had trouble getting comfortable in the train's six-foot long upper bunk. He was 6'4" tall, so he let his feet stick out the end of the curtain. Fortunately, I seldom have trouble sleeping anywhere (but I'm not 6'4"). When the call came for breakfast, I was soon up and ready for the rest of our trip back to our Base.

The lovely Fuji-yama. Photo Wilipedia in the public domain.

Our final Japanese adventure came the

morning we were to depart Japan from NAS Iwakuni for the United States. We had six suitcases and Don and Alice Sumner had eight suitcases, and three young children. The boat from our housing in Niji Mura to Iwakuni left between 5:30 and 6:00 a.m., so we were up before dawn. It was dark and cold when we got to the broken-down sea wall that we had to climb across to get to the boat landing, and our dressy clothes made scrambling quite difficult. When traveling on military airplanes, wives were then required to wear skirts with nylons and proper shoes. One slip and we would be in the Sea of Japan, and it was still dark. Stretch and Don made several trips across the tipped slabs of concrete carrying suitcases and then the children before returning to help Alice and me.

When we got to the crash boat, the coxswain (boat driver) was not there, and we learned that one of the two engines on the boat was inoperable. We waited and waited and finally the coxswain arrived. He radioed the Base and they told him to get underway at slow speed. They had dispatched a second boat from the base to assist us. They also told us that our plane to Tokyo was waiting for our arrival. That plane usually departed on schedule and we were moving at a snail's pace. We met up with the second boat somewhere in the middle of the Inland Sea. The men transferred the fourteen suitcases from one boat to the other while sailors held the boats together. Don climbed over to the other boat, and Stretch handed the children to him while the sailors held the boats as still as possible. The sea was calm, with no big waves, and we were grateful for that. Alice and I must have made quite a sight in our straight skirts and high heels climbing over the railings of the boats, but we made it! We finally arrived at NAS Iwakuni where other members of the Squadron and wives were waiting to say goodbye. We were rushed aboard the waiting plane and were soon on our way to Tokyo.

The following day was Thanksgiving and we made plans to join Don, Alice, and children at the Officers' Club in Yokohama for dinner. They were returning to the States on a ship and we were flying. We had a day before our plane departed and their ship was not leaving for a couple of days. During that evening we received a call that we were now booked on a flight the next day. Without success, Stretch called every place he could think of trying to locate Don and Alice to let them know that we could not meet them for dinner. It was not until weeks later when we arrived in Monterey, CA, that we could explain to them why we did not show up for Thanksgiving dinner.

Flying during those years was on prop driven planes. Flying time from Japan to the United States was approximately 36 hours, with refueling stops at Wake Island and Hawaii. While at Wake Island, base personnel served us a complete Thanksgiving dinner. Our next stop was Hawaii and there we had a second turkey dinner. Soon after departing from Hawaii, a loudspeaker announcement informed us that instead of the usual box lunches we would be served a Thanksgiving dinner. I think many of us groaned at the thought of more turkey with all the fixings. The flight from Hawaii to Travis Air Force Base that day was fourteen hours, but I do not think anyone left that plane hungry.

As our plane took off from Tokyo, I looked back at the area, thinking that I would never see it again. However, we were back in Japan two years later for two wonderful years at NAS Atsugi, which is near Tokyo. A third trip to Japan took us back to MCAS Iwakuni in 1962. The Base had been taken over by the Marines and now had the title of Marine Corps Air Station (MCAS) Iwakuni. That time we lived in Iwakuni before moving aboard the Base, so Stretch did not have to commute to work on a two-hour boat ride across the Inland Sea.

In 1965 a fourth trip across the Pacific Ocean took us back to NAS Sangley Point for two years. It had been ten years since we had left that place. Much had changed, but the swim program that I established was continuing the same way I originally had set it up.

Our life had really changed with the addition of three children. Each time we returned to the States, we adopted another child. My first trip across the Pacific Ocean was by ship, and subsequent trips were by plane, until the last trip in 1967, which was by ship. Stretch said that we had to break the cycle of crossing the Pacific Ocean, by plane. He decided that we would return by ship and not adopt another child. It worked. We did not adopt any more children.

During Stretch's tour at NAS Atsugi, he was the weather forecaster in addition to flying the Station airplanes. While stationed at MCAS Iwakuni the second time, he flew P5M seaplanes in the Navy Patrol Squadron VP-50. The last overseas tour at NAS Sangley Point, he was the Base Maintenance Officer and later Operations Officer. His final tour in the Navy was on COMNAVWESTPAC Staff at NAS North Island, CA. Then he retired and went to work for Lockheed.

I continued my volunteer work for the American Red Cross, establishing Water Safety Programs at overseas bases and also when in the United States. I became a CPR instructor, and instructor trainer in addition to being a First Aid instructor, and instructor trainer. After serving fifty years as a Red Cross volunteer, I retired. I also returned to college to earn a Master's Degree in Recreation Therapy, specializing in Hydrotherapy. I taught all levels of water safety and trained adults to teach water safety and become life guards.

The friends that I made during my first trip to the Philippines and Japan will forever have a special place in my heart. The VW-3 Det.A and VQ-1 tour of duty with my husband was an outstanding and rewarding experience. Writing my story and reading the stories written by the other wives has been a trip down memory lane. It was the "High" point of my life.

Pauly Varner

Stretch & Pauly Verner

Pauly Varner

Autobiography

February 28, 2011

The first trans-Pacific flight landed in Australia in 1928 on the day I was born. Much of my life revolved around airplanes. I met my husband when we were members of the Civil Air Patrol. We married when he was a U.S. Navy pilot. I traveled across the Pacific Ocean six times by plane and twice by ship, in the days before Jet Planes.

I spent fifty years as a Red Cross volunteer teaching swimming, First Aid, and CPR. I did this in the United States and at overseas Bases in the Philippines and Japan.

Later in life, I returned to college and earned my Master's Degree in Recreation Therapy at California State University Northridge. I also earned my teaching credentials and taught at the College of the Canyons in Santa Clarita, CA.

My husband and I adopted three children and I now have eight grandchildren and five great grandchildren. The children are Lise Meyers, Bill Varner (deceased), and Richard Varner. Grandchildren: Katie Gilmore, Cody Johnston, Matthew Meyers, Rebecca Meyers, Ben Meyers, Alana Meyers, William Varner (aka Brian Mikalinis) and Justin Varner (aka Justin Mikalinis). Great-grandchildren: Dominic, Angelo, Lola, Erica, and Amanda Mikalinis.

My husband died in 2005 and I live in San Marcos, CA.

Margaret Burdg

VW-3 DET. A AND VQ-1
Memories of Navy Duty in the Philippines and Japan:
Sept. 1954 – 2009

It was spring of 1954. Sen. Joe McCarthy was deep into his American witch-hunts for communists on American TV and was finally being challenged to face the music for all the careers he'd ruined. In Hutchinson, KS, there were tornado threats and a real dust storm to challenge the air antics of the US Navy officers finishing their training as multi-engine pilots at Naval Air Station (NAS) Hutchinson. Navy training in Kansas? Why? Those vast, dry, and very flat plains don't seem very nautical to the casual observer. But there we were, my husband, Lt.j.g. Don Burdg, and a number of other potential pilots honing their skills as long-range patrol plane pilots soon to fly out over the vastness of the Pacific in search of bad weather. These men and their wives were getting excited about the guys at last getting to wear those gold pilots' wings and head out for one of the Navy's far-flung stations in the Pacific.

When asked for their first choice of a duty station after training, many guys, including Don, listed Barber's Point, Oahu, as their first pick. But when the orders finally came, Don and several others were assigned to a patrol squadron on Guam, a tiny island in the South Pacific, eight miles wide and 25 miles long.

After we all got to pin on our husbands' wings, Don and some others were sent to San Diego for six weeks of CIC (Combat Information Center) training.

First, we had a month of leave, so we drove out to visit relatives and then headed for the west coast.

Those were pre-credit days, and just as we pulled into San Diego and found a motel, we needed some soda crackers to settle my stomach, but we had too much pride to ask for any. So, we realized our budget for the trip hadn't quite planned for enough cash to carry us through. We paid for the motel, but then were completely out of cash. Not even a penny for a gumball. No money for supper or breakfast, no place to cash a check, nothing! The next day Don could go to the Base and get some back pay owed him, but that wouldn't be available till 4 PM. I was pregnant, with morning sickness. I badly toughed it out until 1600. Of course, I felt compelled to tell Don that my dad, whom we'd just visited, would never have let Don marry me had he known I'd be starving and pregnant, with no money, thousands of miles from dear old Colorado.

Finally Don made contact with the paymaster, and took me out to dinner. I remember I ordered the most expensive thing on the menu. Can't remember what it was, but I know it cost $10.00.

After CIC, Don was ordered to leave soon for Guam by plane. I, of course, would have to go by ship. Don would depart from San Francisco, and he had about a week of leave time, so we drove there to explore new territory before he had to depart. What fun! We discovered some wonderful restaurants and walked through famous stores, such as Gump's ("where good taste costs no more"). We walked and drove up hill and down, sampling the city and finding it exciting, magical.

But soon it was time for Don to fly to Guam. I sadly kissed him goodbye, drove our car to the Base from which it would be shipped to Guam, and took a Greyhound bus to LA, where my sister was doing an Occupational Therapy summer internship. She'd rented an apartment there, which we were going to share until I got my sailing orders to head for Guam and Don.

Well, the bus trip from San Francisco to LA was very uncomfortable. Every seat was taken, and my seatmate was a Marine who promptly fell asleep with his head on my shoulder. I couldn't rouse him, so I just sat there stiff and missing Don, until we pulled into LA. The LA bus station in 1954 was daunting!

There were very scroungy-looking people everywhere. I was a little country girl who'd never been in a place like that before. I was scared!

My sister hadn't met me because she had no car, so I had to find a cab and go to her apartment. Again, I felt terrified. Just knew the cab driver would try to rob me or drive around blocks to run up the fare beyond what I had in my purse. But, it all turned out okay. Our tiny furnished apartment was kind of cute, safe, and near a city bus stop. This gave me access to various Navy facilities where I could get all the immunizations and paperwork needed for my trip to Guam. During a pre-natal exam, a Navy doctor told me I'd have to get a passport and be ready to go to Guam by Sept. 1 or I'd have to wait until after the baby came. To expedite matters, I took a one-day round trip flight to San Francisco and through the State Department office there, got my passport in time.

Well, Don's first letter brought some very puzzling news. He's arrived at Guam, checked in and then been notified that a detachment of a squadron in "RoP" needed a pilot and he'd been given the choice of either going there or staying there. He said he'd decided to go, thinking life in that country would be far more interesting than living on the tiny isolated island of Guam. So, he was doing the paper work to change my sailing orders. Well, I had no idea where "RoP' was! We had no encyclopedia in our apartment, and neither my sister nor I knew where a library was. This was eons before e-mail, the web, and even long before overseas phone calls. I didn't know what to think or do! Don's address was FPO (Fleet Post Office) and then some numbers, so I wrote to him and soon got clarification: I'd be going to the Republic of the Philippines! I didn't know where that was, except in Southeast Asia. But my bioclock was ticking, growing our baby, so I shifted into high gear and got ready to go.

My sailing date was Tuesday, Sept. 7, from San Francisco, the day after Labor Day. I took my sister with me to SF, and we did my last weekend in the States for two years by getting better acquainted with Frisco, a place she'd never been. It was fun! She especially enjoyed China Town. Sunday she had to fly back to LA and I went back to my lonely hotel room and waited for Tuesday.

I can't remember getting on the ship, the USNS *Fred Ainsworth*, but I do remember my tiny cabin with bunks for four. My roommates were two other young junior officers' wives. One was pregnant, like me, and a Navy wife, too, but she was going to Guam. The other was an Air Force wife, going also to the Philippines, but to Clark Air Force Base (AFB), not to NAS Sangley Point like I was. We got along great . . . a good thing, because our journey was to take 23 days.

The first two days of our trip, the seas were rough and many people got sea-

USNS "Fred C. Ainsworth" (T-AP-181) off San Francisco, CA., 11 August 1955.

sick. But the wide, wide Pacific soon settled down, true to its name, and we had a smooth, *long* ride. Our days were filled with food, nearly round the clock eating opportunities. I gained 10 pounds in three weeks, and the first Navy doctor I saw at NAS Sangley Point had a fit!

I remember our days on board were also filled with bridge games. I'd just learned to play bridge that spring and was scared to bid. I usually was playing with wives of officers who far outranked Don. These women were gracious, but older, more mature, and I felt outranked in every respect. I'd pass, even with 22 points (with only 14 you're required to make an opening bid!) and soon incurred the wrath of so many partners that people quit asking me to play. So I settled down to reading in a deck chair. Got the best suntan of my life! When we crossed the International Date Line, most of us were dunked in a tank on deck, but I was spared because of being pregnant.

22 days after our journey began we docked at Guam. As we pulled in, I couldn't believe my eyes! There was my husband, Don, standing on the dock waving with the rest! Turns out his Executive Officer had let him be part of the crew on a squadron mission to Guam that day, and we got to spend the day together. Even visited friends we'd finished training with at Hutchinson just a few months before! Of course at day's end I had to get back on my ship, and he had to fly back to NAS Sangley Point. When his plane flew over us, it rocked its wings, and I was awed by my first encounter with the powerful, beautiful P4M-! Q!

A day later we docked at Manila, and there was Don again! After a short trip across the bay in a much smaller boat than the *Ainsworth*, we docked at NAS Sangley Point, our new home. Some VW-3 Det.A Squadron wives met us, put a wonderfully fragrant lei of white flowers around my neck, and took me to one of their homes on the base. Don left to get my luggage squared away, and the other wives and I shared cookies and iced tea, while I enjoyed looking at the cool elegance of the base home. I couldn't wait till we got one too. But Don had already told me it could be at least a year before we could move up the housing list far enough to get base housing. In the meantime, we'd have to rent a Filipino house in Cavite, the Filipino city adjacent to the base. The squadron wives gave us some leads, and their sincere best wishes, and I remember their warm, sincere cordiality. This started a bond of loving concern between us which we all (husbands too) depended on, through thick and thin, in our lives in this squadron. Don and I still share and enjoy them over 50 years later.

But I'm getting a little ahead of myself. After that refreshing welcome, Don picked me up and we headed for the house he'd rented for us in Cavite. The minute we stepped outside that air-conditioned base house, the very humid, very hot tropical air hit me. Gone was the sea breeze we'd had on the ship. Cavite's sticky heat was like Pensacola's, where we'd lived for a year while Don was taking basic flight training. But Cavite's was different, because as soon as we got outside the gate, there were new, pungent odors to get used to: very strong, fishy smells, both raw and frying, and always, the thick, unappetizing odor of frying garlic.

Since our car hadn't arrived yet, we hired a "jeepney." The Filipino version of a cab. It was an American jeep restyled, with a colorful top and side benches rather than seats. It cost us 10 centavos (about 5 cents) to take us to the house. There was annoying, loud music playing next door, and it followed us inside, because the windows were wide open (always were, I soon discovered, to let in any moving air). I can't really remember anything about that place except the bed. I was very tired from the long day and from being very pregnant. With swollen feet, so we got into bed. We learned later that it was stuffed with coconut husks, and was standard for the ordinary Filipino home! Well, we both tried and tried to get to sleep. But, it was so hot you didn't want even a sheet over you. The crunchy mattress was bizarrely uncomfortable, and the annoying loud music never stopped and couldn't be shut out because closing the windows made the room stifling. Don apologized for this place and said he'd see what he could do about finding another place the next day at the squadron.

Well, we were in luck! Another squadron family, the Van Allens, had just qualified for base housing. They'd been living in a house in a nicer residential area of Cavite, and the neighbors were not noisy. We quickly went to check it out. This house had some attractive rattan living room furniture and three bedrooms, with nice built-ins in the master. The kitchen had a stove but no fridge (the Van Allens had bought one from the base Navy Exchange, and taken it with them to their new quarters) There was a nice looking double bed, but it also had a coconut husk mattress! But again, Van Allens said, we could get a good American one from the Navy Exchange, along with a fridge, in short order. There was safe running water (no hot), but a nice tiled bath with shower, which, we soon learned, often only dribbled tepid water until you yelled "Tubig-sa" (water please), to let the neighbors on the same water main know your needs. There was even a fenced-in yard, with a huge banana tree in the back. We checked everything out and quickly decided to rent that house. We made one bedroom into a nursery for our soon-to-be-born baby, and the third bedroom we gave to our house girl, a very capable, gracious, and honest Filipina named Irenea Enrique (Irene).

House girl? Yes, everyone in the squadron had told us to hire a house girl as quickly as possible to help us keep house in the tropic. The Navy even gave personnel a monthly allowance for paying household help. No house outside the base had a washing machine or dryer or clothes lines. Your house girl did your laundry out in your front yard in the boiling hot sun, squatting down on her haunches, her skirt tucked in around her thighs. We came to call this position the Asian squat, because we saw it everywhere. It was how people did business, played games, whatever. The tepid wash water came from a front-yard hydrant, and the wash and rinse "cycles" were hand done in two big fluted tin pans. After being scrubbed, rinsed, and wrung out by hand, the laundry was then draped on the yard bushes and dried — whites in the sun, coloreds in the shade. Every day all the wash was neatly folded or ironed and put away by Irene.

The tropical heat affected everyone on the base. Office hours were from 0700 till 1300. The all-male military personnel went to work every day in tropical uniforms. For the officers: khaki shorts, and short-sleeved, open-collared khaki shirts, and caps rather than hats; for the enlisted men, dungarees, blue baseball caps and blue cotton shirts. All uniforms had to be washed and ironed frequently. This was before Permanent Press! Our bottom bed sheet needed to be changed everyday because our house wasn't air conditioned, and every morning there was a sweat print of where we'd slept. Humidity was at least 90% every day, and the temperature always seemed hot. We had to keep our clothes and small appliances in closets called hot lockers, which had light bulbs burning in them at all times. Otherwise, clothing, including shoes, belts, handbags, plus anything metal, such as a toaster, would mold or corrode.

However, there was a plus side to this hot sticky climate. We could buy wonderful fresh fruits and vegetables very cheaply in Cavite: e,g, a kilo of either tomatoes or lots of different kinds of bananas cost only 5 cents! But all fresh produce had to be soaked four hours in a solution of Clorox and water to kill dangerous bacteria from the "night soil" (human feces) used as fertilizer. Then, before we ate these deliciously fresh things, we rinsed them in more water to kill the taste of the Clorox. Our house girls did this time-consuming job for us.

Two more definite pluses to our Filipino life were our landlords, Mr. and Mrs. Victa. They were a very nice middle-class Filipino couple. He was a civilian worker on the base and she was a home economics teacher in a Cavite school. They warmly welcomed us to their country and city and gave us some words of wisdom to help keep us safe. They told us to always keep our doors locked at night or when out of the house to protect ourselves from "cat burglars" who had the habit of sneaking into "rich" American homes barefoot to steal things, even wallets, if they'd been left in pants draped on a bedpost overnight.

The Victas also warned us to be very careful if we ever drove out into the countryside, especially to Manila. They said if we ever happened to hit or injure a local native, to drive away immediately, before a crown could gather. Our landlord looked us straight in the eyes and said, "They (the locals) will have to kill you!" We couldn't believe it, but they both insisted it was true. They said there were no police in the countryside and that this was the accepted way of meting out justice! Well, we followed their advice and took a regularly scheduled Navy boat across the bay when we wanted to go to Manila and back.

However, we did want to do a little sightseeing to places we'd heard about that were on safe roads. Our car finally arrived from the States, and one Sunday we and our next door neighbors packed a fried chicken, potato-salad type picnic lunch and ventured out to see the famous Lake Taal phenomenon: a tiny island in a lake in a volcano on an island (Luzon) in the sea. After seeing this and taking some pictures, we found a nice grassy spot to sit on, got out our picnic lunch and started to eat.

Within seconds we were surrounded by about ten kids, barefoot and clothed in rags, who squatted down on their haunches and stared at every bite of food we put into our mouths. We looked at each other and decided, without a conference, that we should pack up our picnic and leave. It was uncomfortable to eat in front of those kids, and yet if we'd left the food for them, we knew they'd probably get into a big fight. So that was our last Sunday picnic outside the base.

Don had been given some leave time for my arrival and finding housing. But soon it was time for him to be back on the squadron's mission schedule. I quickly learned that our squadron was tightly wrapped in Top Secret security. The planes were painted black and had no insignia was on them. The missions were always flown at night. This meant that Don would come home from work right after lunch and take a nap. Then at dusk I'd drive him and his flight bag to the base and leave him at the BOQ. Sometime that night he'd go out on a mission. He could never tell me where he was going or when he'd be back. It might be the next morning or several mornings later. All the crews would only describe where they'd been as "Up North."

The other squadron wives were quick to include me in "girl things" we could always do when our guys were gone. We never mentioned the missions. I did sense they were dangerous, but also knew the only thing I could do about that was pray. Which I always did. Meanwhile we wives might go to Manila to shop for new material, which we then took, with a magazine picture of a dress we wanted, to a Cavite dressmaker. In less than a week, for a very small price, we'd pick up the finished product! We all lived in sundresses, shorts, and swimsuits. I did learn to play better bridge and enjoyed having lunch or dinner with other crew wives at the O'Club and then taking in a movie in the outdoor base theater, where the evening air was still humid, but not so hot. Once in a while all the squadron officers were home and then we'd have a great party at someone's home. There were cocktails and good food and lots of laughing and talking, with the guys telling plane tales with their hands in one corner and the gals sharing news of good buys and base gossip in another.

Although they never said where "Up North" specifically was, the guys did brag about have wonderful langouste (like lobsters, I guess) dinners there for $1.50. We were so jealous that one time squadron wife Mary Jane Edixon and I talked our guys into taking us to the Manila O'Club and we ordered lobster dinners: $10 each, which was a fortune in the 50's! Well, the cheapskate guys countered by ordering hamburgers for themselves!

Soon it was Thanksgiving. We were invited to dinner at the Van Allens' new place on the base. We had a wonderful turkey dinner with all the trimmings. I LOVE food, and I stuffed myself. I can still see my self, 8½ months pregnant, laid back in a comfortable rattan chair with my feet up on a foot stool, happy as a clam!

Maybe it was eating all that food or just time for our long-awaited baby to come, but the next night I went into labor. Fortunately, Don was home, not out on a mission. He drove me into the base infirmary (there was no hospital) and our second son was quickly and safely born, on Nov. 29, 1954. We"d lost our first son at Christmas time the year before in Pensacola. He had been born at 6½ months, and, like President and Mrs. Kennedy's third child ten years later, didn't have sufficiently developed lungs to survive.

But this baby looked wonderful! In face, he looked like Don, who's quite a handsome guy! We named him Mark Austin, (his middle name was my dad's). We wired our families and were so thrilled! I nursed and was looking forward to taking him home soon. Don took pictures of him in the nursery.

But when Mark was two days old, a doctor came early in the morning and told me Mark was becoming dangerously jaundiced, a reaction to my Rh negative blood antibodies, and needed an immediate blood transfusion of Rh positive blood. They couldn't do that at our base, so they flew Mark and me by helicopter to the hospital at Clark Air Force Base, 30 minutes away. They let Don come too. And we prayed and waited. Later that afternoon a chaplain came to tell us that Mark had died. We couldn't believe it! He'd looked so healthy at birth. But evidently he had contracted pneumonia in the infirmary at Sangley Point and could not fight that and the Rh jaundice too. Why the infirmary personnel had not sent him to Clark AFB sooner we were never able to find out. His death seemed so preventable to us and to all our squadron mates back at Sangley. They all rallied to comfort and help us get beyond the deep grief we felt.

But, life goes on. We still had a year and a half tour of duty to serve and had to make the best of it. It was the Christmas season, and all the Filipinos put huge colored paper star decorations on their front porches with streamers waving in the breeze. Groups of caroling musicians walked the streets at night, like mariachi bands in Mexico, playing the old traditional carols with a Latin swing, and singing the words in their native Tagalog. We would go out and give them a few centavos in appreciation.

We received a unique Christmas present that year: a Filipino puppy, just about a month old. She was given to us by a couple I'd traveled with across the sea to the Philippines. Of course this gift was meant to cheer us up, and for the rest of our stay in the Philippines this pup certainly kept us occupied and entertained. At first she cried all night, but after we put an alarm clock with a loud tick in her bed, she settled down. Evidently the ticking reminded her of her mother's heartbeat. She licked and licked the clocked and pretty soon didn't need it.

She grew very fast — looked like a black and brown terrier with long legs. We named her Zigi-na, which means Go fast! She was always on the go. She liked to jump up on us, but only when we went outside to get into our car. Our driveway was usually muddy, so she left paw prints on our clothes, especially annoying when we had just enough time to get wherever we were going *without* an unplanned change of clothes. But finally we discovered a solution for this problem. Zigi was scared of water. Hold a glass of water over her head and she'd cower. So we kept a glass of water on a fence support just inside our gate and used it to keep Zigi down. She did bark at all strangers, which we didn't mind, because we figured she'd scare away potential burglars. This worked, we were never robbed. Guess it was because no potential criminals ever discovered the "water torture,"

Since many Filipino dogs were unhealthy, we never let Zigi out of our fenced-in yard. So, she'd never been around other dogs. One Sunday we had another squadron couple over, and they brought their dog, which was much smaller than Zigi. The little guest came up to "meet" Zigi with a friendly sniff, but Zigi jumped back in fear. This happened several more times, but Zigi finally accepted the "guest," and they eventually romped in the yard together. When it came time for us to leave the P.I., we had to find a new home for our dog. Our dear house girl, Irene, agreed to take her and found a good new home for her.

Another distraction that helped us get over the loss of our baby was tennis. Don, a good tennis player, had qualified to be on the NAS Sangley Point tennis team, which was invited to participate in an inter-service tournament with an Air Force team. The wives got to go along, and it was such a treat! The tournament was at a lovely place, high in the mountains north of Manila on a road safe for driving, called Camp John Hay.

This place was a very popular R&R destination for military personnel and their families to enjoy leave. The climate was cool, and there were very nice cabins with blankets on the beds. What a blessed reprieve from the heat and humidity of Cavite! This camp was right next to the Filipino city of Baguio. Some very interesting natives called Igorots lived in or near the city. They did authentic folk dances for tourists and also hand carved and sold beautiful wood carvings. We bought a figure of a dancing Igorot woman, a bust of a handsome Igorot man, and a beautiful salad set, which we still use.

Of course this trip made us feel good, and one incident in particular was very good for me. It helped me get over the lingering heartache from Baby Mark's death. I had a glass of wine with dinner at Camp John Hay. After drinking the wine, I noticed that the deep pain in my heart was gone. The wine had taken it away! But, I also realized that there could be real danger in drinking alcohol to relieve emotional pain.

Bamboo bridge across river on the way up the mountain to Baguio

So, I went back to NAS Sangley Point with a new attitude. I could tell Don was very involved in his squadron work, even though he couldn't talk about it, and I wanted to find something positive to do as well. I believe our Higher

Power gives us little nudges when we need them, and right after returning to our base, I got a good nudge. The school officer of NAS Sangley Point contacted me to see if I'd come in for an interview. I hadn't applied for a base teaching job because I wasn't an elementary teacher. Besides, I'd thought I'd be too busy to work and take care of a new baby too. But the school officer was so cordial and insistent that I went. Somehow, the word had got around that I was a certified high school teacher. True, the base school was elementary, but there was a need for someone to finish the year teaching third grade. That teacher was very pregnant and her husband had orders to go Stateside, so they were leaving very soon and someone was needed to take her place. Well, to make a long story short, they hired me.

That job was my salvation, I had 20 third-graders. Our school was a series of Quonset huts with no air conditioning, but tiny fans in each corner kept the air moving. The school's tropical hours were 0730-1200, with a 20 min. snack recess at 1030. Although the kids had loved their teacher, Dottie Moss, they readily accepted me and together we finished the year very positively. They worked hard, played hard, and were just great kids. Each one came from a two-parent family, and all the parents supported me. In addition, all the teachers and principal were very cordial and helpful, and I was thrilled to be asked to teach third grade again the next year.

During the school's summer break another squadron wife, Alice Sumner, and I went on a trip to Hong Kong. By then both of us had seen lots of goodies brought back from there by other people at our base, and we wanted to get in on some of this fabulous shopping ourselves.

Causeway Bay in Hong Kong in 1955. Pubic domain image courtesy Wikipedia.

Don had already been to HK with two other pilots from the squadron. I couldn't go because I was teaching, but I was really impressed by the pictures Don had taken and by what he'd bought. Which were some good-looking tailor-made sport coats and slacks, and two English bicycles. I needed some new winter clothes, too, and wanted to shop for a number of things that friends at Sangley had purchased there, such as jewelry, fruit made of semi-precious stones, embroidered linens, and beautiful copper and brass items. Also, I just wanted to see and shop in this international mecca.

Alice's husband had already been there, too, so she and I went together. As I signed the hotel register, I noticed the writing, in English, on the line above mine. His name I can't remember, but his country of origin was the USSR, his citizenship: Russian, his occupation: engineering, and his destination: Peking (now Beijing) via train. We were staying in a hotel with a Russian communist! Unthinkable in the mindset of our times! I was stunned and suddenly realized that we were truly on a cosmopolitan adventure and could encounter, without even realizing it, people from Cold War countries behind the Iron Curtain, countries to which no Free World people like us could travel to in the 1950's.

Eating in the hotel restaurant, we heard only French from a family at the next table. And we soon learned that everywhere we went we would hear snippets of many other languages. One of the first shops we went into had lovely pieces from Czechoslovakia (now the Czech Republic). Then these items could not have been found anywhere in the U.S. Alice and I soon realized that being in Hong Kong was like being everywhere in the world at once.

Once we were settled in our room, we decided to go on our first shopping excursion, to a jeweler's named Ada Lum. We'd admired many of her handcrafted pieces on women at NAS Sangley Point. We had her address and learned how to access the famous Hong Kong red, double decker bus system to get there. (We didn't want to spend the money for a cab.) It was late afternoon when we started out. We tried to find out from our bus driver where to go after we got off the bus, but we soon found ourselves walking through strange-looking groups of people cooking food right on the sidewalk. We also saw some people sleeping on the sidewalk, near the cooking fires. It was after dark when we finally found Ada Lum's address. The person answering the door left us outside and went to get Ada. I'll never forget her commanding presence as she invited us inside. She was tall and statuesque. Her hair was swept up, and she was wearing a black cheong sam, the traditional sheathlike dress of Chinese women, with a high neck and a side seam slit to the thigh. She also wore heavy gold dangling earrings.

She asked us how we had found her. When she learned where we mistakenly walked, she immediately scolded us for taking such chances. Evidently our route had taken us through a very dangerous settlement where homeless people lived right on the street. Possibly they were refugees who had illegally escaped from Communist China. There was not enough housing for such folk in Hong Kong, so they lived on the street and preyed upon any passerby who might have money. Ada Lum told us we could have been killed and that we had to take a taxi back.

Having said that, she immediately morphed into the purveyor of the famous jewelry our friends had bought, and we did find some nice earrings (the only things we could afford).

Our main shopping goal for this trip was to have some new winter clothes made. We found the tailor our friends had recommended, Lee Kee, and proceeded to pick out fabrics. I ordered a warm gray winter coat, a creamy cashmere winter jacket, and a blue serge suit. Alice picked things too. As soon as our orders had been placed, we were extensively measured and told the clothes would be ready in two days. At a nearby shop, I also ordered a black suede handbag and matching pumps.

Next we focused on sightseeing and eating. We took a night tram to the top of Victoria Peak and looked down on the famous view of Kowloon at night, exquisite — as someone has said, like sparkling jewels scattered on black velvet. In contrast to that, we endured the garish Tiger Balm Gardens. To us, sampling different cuisines became much more interesting than additional sightseeing. We really enjoyed a Russian restaurant, where we had borscht and "black bread," absolutely delicious.

In spite of all the fun we were having, one afternoon Alice and I found ourselves involved in a frightening incident. All of a sudden on a crowded street, people began shouting and looking up the street. A man came running down the street shouting very loud words we couldn't understand. Finally, we were able to find an English-speaking store clerk to tell us what was happening.

Evidently the man running had been robbed by a pickpocket and was trying to chase him down before the culprit was lost in the crowd. That was it. But afterwards Alice and I confessed to each other that all that shouting and those worried faces had made us both wonder whether the Communist Chinese were at last really invading Hong Kong, then a British colony, but economically and ethnically very attached to Communist China. Our fears might seem foolish now, but those years, 1954-56, were Cold War years, and we in the Free World were always wondering what our enemies behind the Iron Curtain would do next.

Our last night in Hong Kong Alice and I were treated to a French dinner by a friend from our squadron who happened to be in Hong Kong, too. Eating at a well-known restaurant called Gaddi's, we ordered turtle soup, and something else; to top it off, we had the most extraordinary dessert I've ever savored: flaming crepe suzettes. This was prepared at our table. First the tiny crepes were baked and kept warm while the delicate sauce was made, combining butter, brown sugar, and splashes of nine different liqueurs, After all these ingredients were perfectly blended, the sauce was briefly set aflame, then served over the crepes. Today we'd say it was to die for, but then, and still today, it is the most delicious dessert I've ever tasted. To preserve that memory, I've never had it again.

The next day we went to tailor Lee Kee's to pick up our orders. After trying the clothes and shoes on, we watched clerks pack our purchases in little brown cardboard suitcases boldly labeled, "Lee Kee, Tailor, Hong Kong." Then we went back to the Russian restaurant and bought loaves of black bread to take home. Soon it was time to catch our plane home to the Philippines. We'll never forget that unique experience of being everywhere in the world at once.

When the next school year started, I had a wonderful new group of students. They were "my kids" and we were traveling a good learning path together. There was only one problem: I was pregnant again — certainly not by choice this time. This was before "The Pill," and the infirmary offered no contraceptives. We had definitely wanted to wait to try having a baby again until we returned to the States and could consult doctors about using a drug we'd heard of that could neutralize the deadly Rh Negative antibodies which caused my blood to fight with Don's.

These antibodies increase with every pregnancy and can poison the baby at birth, unless the child has an O Positive blood transfusion. This had been the main cause of Mark's death. I shared my great concern about this with the Sangley Infirmary personnel, and they tried to encourage me, but could only offer to send me to Clark AFB for delivery of this my third pregnancy. However, that didn't happen because this baby was stillborn at six months.

Don's CO told him we could go right home to the US and be relieved of his remaining duty time because of the hardship of losing three babies in three years. But, rather than accept that offer, we chose instead to join our whole squadron in moving to NAS Iwakuni in Japan. Earlier that year our squadron, which had been Detachment Able to Navy weather squadron VW3 for years, had finally been commissioned as a full squadron, named VQ-1, Electronic Countermeasures Squadron One. I didn't understand what that meant, and Don couldn't tell me, but I did get a hint that the truth was classified after we moved to Japan when we went to a welcoming reception at NAS Iwakuni's O' Club. The table décor included lots of question marks and the word Shh!, implying secrecy.

So we wives had finally learned that at least part of "Up North" was Japan. And Don, it turned out, had already flown many missions to and from there and liked the country and its culture. From "Up North" he'd brought home a pretty set of Japanese china, a nice brass lamp, and a lovely string of pearls to me. Now he was eager to have me share in discovering more of Japan with him in the seven months we had remaining in our tour of duty.

So, I regretfully resigned from my job and said fond goodbyes to my third graders. Some of them even wrote to me after I got to Japan. We sold our little "one-eyed" 1937 Chevie coupe that had been our transportation in Cavite after we'd sold our nice '53 Ford to the Filipinos. And, when the Navy movers came to pack our household gear for shipment to Japan, we also had them pack the new English bikes we'd bought in Hong Kong. Those bikes would be our transportation in Japan.

The next months were the best part of our Far East tour. We got to leave our hot home in Cavite and move into *base housing* at NAS Iwakuni, located on the southeastern shore of Japan's largest island, Honshu. It was fall, the air was cool and crisp, and the leaves were changing to beautiful autumn tones. There was one problem: I had no winter clothes. The sun dresses I'd had made in the Philippines just didn't cut Japan's cool autumn climate. I had asked my dad to send some of my old winter duds, from Colorado, but they had to be sent by ship, which could take months. So dear squadron wife and friend Joan Angier loaned me some of her warm clothes till mine arrived. She and husband Don had preceded us to Japan and their home on the base was very close to ours. What sweeties they and their little kids, John and Ann, were!

Of course Don was regularly flying night missions from NAS Iwakuni, but when he was home, we went for great bicycle rides out into Iwakuni. The air was often cold, but we bundled up in the new wool coats we'd had tailor made in Hong Kong and enjoyed exploring the interesting landscape and quaint town of Iwakuni. The stores were unheated except for a large coal-burning open pot in the rear. The store clerks, all friendly, wore layered kimonos and footlets to keep their feet warm. These had the big toe separated so that the wearers could slide into their Japanese versions of what we now call flip-flops

There were many neat things to buy, especially many varieties of attractive pottery and lacquer ware. If we seemed to be cold, shopkeepers would invite us to come in back and warm our hands near their heaters. We bought huge blue wire baskets for the backs of our bikes to carry home the treasures we'd buy. One Sunday we carried this to the extreme: we bought a whole set-eight place settings, plus serving dishes — of pottery to take home to the States. Believe it or not, we filled both bike baskets with the packed boxes of pottery and rode our bikes safely back to the base, with nary a chip in our new dishes!

At night we really enjoyed eating in charming little Japanese restaurants in Iwakuni. Softly glowing, hanging paper lanterns outside a restaurant would be the only advertizing: no neon. Just walking by the open door, you'd whiff wonderfully appetizing smells of spices, tea, and tasty meat and vegetable dishes made with savory sauces and served on rice or noodles. They tasted even better than they smelled, especially if you ate them with chopsticks. Handling these was awkward at first, but we soon learned the knack of it and still enjoy using them when we have oriental food. The charm of Iwakuni at night is probably all changed now, fifty-five years later. But in my mind's eye I can still see, smell and taste those Japanese delights.

All the Japanese people seemed friendly. We learned a few phrases of their language, and it pleased them for us to use it. But if we started to struggle, they'd speak some broken English to help us out. We weren't far from Hiroshima, a city completely destroyed by our A-bomb attacks in 1945. It was only 1955, ten years later, when we visited there and were amazed that the city had been completely rebuilt, with the exception of the area around Ground Zero. There a museum had been built, and it housed some very grisly exhibits: pictures of the flattened city and of horribly injured and dead victims. There were even some preserved strips of hideously burned skin. Don had been advised not to wear his pilot's wings if he went there in uniform, for that could be a painful reminder to the civilians of who had attacked them. But outside the museum was a low stone arch with an eternal flame burning incense and words of hope for world peace inscribed in the stone.

Christmas in Iwakuni was fun. Our Christmas tree was a joke. The only available trees at the Navy Exchange had been cut in a faraway place months before. As soon as they were decorated, they gave up their futile efforts to look and smell traditional and went limp, shedding all their needles in pitiful piles at their feet. But we enjoyed the holiday anyway. I remember having a new squadron pilot over for dinner. He fascinated us with news of what was happening back home in the USA.

I especially remember his description of the newest rage in the States, a phenomenon in California called Disneyland. We knew we wanted to go there! That holiday season it actually snowed in Iwakuni! Evidently quite a rare event. It seemed surreal to see snow-capped palm trees on the base!

Of course most Japanese don't celebrate Christmas. Their main holiday is New Year's. They decorate by putting tangerines, which they call *mikans*, and fresh pine boughs in swags over their front doors. Part of their celebration is to go to a Shinto shrine to worship. The huge crowds at the shrines were amazing. By the mid-fifties, most Japanese were wearing western clothes, with elderly people being the exception, still preferring kimonos. But for New Year's, many women and girls and some men wore their best, often very beautiful or handsome kimonos. Some women and girls did their hair up with lacquer, and some even wore the traditional white makeup.

Our squadron had a wonderful New Year's Eve party at the O'Club. Many of us gals wore new dresses made by local dressmakers. Mine was blue velveteen, and I loved it until one honest wife told me I was looking very staid. Okay, so I'm frumpy, but I still had a great time that night! We'd made a number of close friends at the squadron, and they all seemed to be there, having fun, sparkly eyed, eating good food, and drinking toasts at midnight, and showers of confetti and waves of horn tooting.

Well, the next day was January 1, 1956, and we knew we had only five more months of our tour. So we tried to take in all of Japan that we could. We'd come to love this duty station. I took a flower arranging class from a charming teacher named Kikui-san. I so enjoyed learning how to use graceful twigs and a few flowers and leaves to create a beautiful yet simple work of art. I still enjoy doing this, decades later.

Kimono clad dancers singing and clapping

One day as a bonus lesson in Japanese culture, Kikui-san invited our class to a demonstration of the traditional Japanese tea ceremony the next Sunday. Don and I were the only ones to accept this invitation, but we were glad we did. It was a unique experience to join a kimono-clad group of women kneeling on thick cushions and having Kikui-san whisper to us the historic symbolism of each gesture in the ceremony. I like the green tea we were served in little round cups with no handles and accepted what looked like a bar cookie to go with my tea. Well, imagine, if you can, eating a bar of chopped up alfalfa, sickeningly sweet, and very sticky. I tried to be gracious and smile while I ate it, but the snapshot Don took of me belied my inner revulsion. Of course taking that shot was his excuse for not trying one of those "goodies" too.

Not long after that Don took a week's leave and we took an overnight train trip to Tokyo. Sleeping in a Japanese Pullman car was quite an experience. While the porters were making up the beds at one end of our car, the middle-aged man right across the aisle, wearing a suit and tie, stood up, put on a kimono, stepped into the aisle and proceeded to undress himself under the kimono! His seat mate, probably his girl friend, took each article of clothing as he handed it to her and folded it neatly away in a bag. As he undressed right at my shoulder, he spoke angry-sounding, very guttural Japanese, casting his eyes in my direction. Then he climbed into his berth and closed the curtains. We figured he might have vented his feelings to humiliate me because I was sniffling with a cold and wasn't wearing the traditional operating-room type mask which all people are supposed to wear to keep from spreading germs in Japan, actually a good idea.

Tokyo Ginza 1955. Copyright Michael Rogge.

Tokyo was interesting and fun. There were huge neon signs on many buildings, and what a variety of restaurants we found! Our favorite we found quite by accident. We were in a cab trying to find a good restaurant we'd heard about, and our driver probably couldn't understand our Japanese pronunciation of its name. He pulled into a parking lot to turn around. Off to the side of the lot was a sign in English that said "Irene's Hungaria." There were steps leading down into it. We were hungry and decided to give it a try. The ambience, food and music were suburb, well worth getting lost for!

One day of our Tokyo trip we went to Nikko, the site of some famous Shinto temples lavishly decorated in black, red, orange, and lots of gold. It was cold, and taking off our shoes to enter each shrine wasn't fun. The one Nikko image that sticks with me now is a beautiful, large fir tree with hundreds of tiny white prayer papers fluttering on its branches.

On our train ride back to Tokyo, something strange happened. The train suddenly stopped in the countryside. After a bit, an employee came through our car, sliced his index finger across his throat and said what sounded like, "Kitura." Some English-speaking passengers told us that evidently a man had thrown himself under the moving train and been killed. We were shocked, not only to hear that news, but also to see the rather stoic reactions from the other passengers. Did that mean this was a common occurrence? We didn't know, and didn't know enough Japanese to find out.

The most interesting entertainment we found in Tokyo was an afternoon at the Kabuki Theater. The whole performance was said to be five hours long, we only stayed about two, but in my mind's eye and ear there are still vibrant images and sounds. We were seated in the back of the theater's large auditorium. In front of us was a sea of black-haired heads: the audience. Then I remember the stage set, sparsely but elegantly furnished only with stylized cherry trees and strings of cherry blossoms hung like beads from the high proscenium arch. The actors wore white face makeup with colored details then added to match the personalities of the characters they were portraying. They all wore traditional costumes from far-gone times. All of the actors were men, even those playing women's parts. The ranges of their voices were amazing from very high, submissive-sounding female to very low, gruff and powerful male. There was a sword fight to the death between two warriors, with all the movements and voices very intense but stylized into slow motion. Accompanying all the action was plaintive string music played by kimomo-clad musicians sitting on cushions, stage left, outside the scenes.

Our last outing of the Tokyo trip was a relaxing couple of days and nights at nearby Lake Chuzengi in a cozy, western-style inn. This meant the building was heated, didn't have the traditional sliding paper walls, and had American-style beds rather than large cushions (futons) on the floor, topped with a thick quilt but no pillow. We should have gone to a traditional Japanese hotel, just for the experience, but our cozy bed with legs and pillows was very comfortable.

Spring in Iwakuni was lovely. It was getting warm, and the huge blossoming cherry trees were exquisite. One Sunday we went into Iwakuni with our camera just to try to capture some of the feeling of this season. Lots of people were out and about walking, crossing beautiful Kintai bridge, with its curving arches spanning the stream. Many people were sitting on the ground on blankets. We saw whole families, and sometimes a whole circle of men in suits and ties. There were many couples, some young, some middle-aged, and some elderly. All these people were having some refreshments: drinking sake or tea, and eating. Many of the kids had the Japanese counterpart of American popcorn. This treat was a small, dried, fried, or baked squid, seasoned with soy sauce and speared on a stick, like a candied apple in the US. The kids sucked and pulled on its tentacles, one at a time, and gradually softened them up enough to chew and swallow. We saw kids with these everywhere — a funny sight to both of us (amusing and odd), but obviously enjoyed. We each thought about trying one but never did.

Don loves taking pictures, and found lots of good potential subjects in Iwakuni out enjoying spring. He'd learned that if he asked in Japanese, "May I take your picture?" most people were pleased to let him snap theirs. This was just about all the Japanese Don knew, but that and "Thank you very much" worked well. And we still enjoy looking at those pics. They bring back that delightful spring along with the sad realization that soon we would be getting orders to leave Japan.

Our friends Don and Joan Angier had just received their orders. Joan had been teaching the base high school students for several months. A replacement was needed to finish the school year, and she talked me into taking her place as THE ONLY TEACHER IN THAT SCHOOL! She'd done a great job, and I tried till we got our orders, but that brief interval nearly turned me totally off to teaching. Why? Because I had to try to cover all subjects! I could handle the English and history, my majors in college, but the physics, math, and chemistry I just could not do! I felt totally over challenged and inadequate. There were only about 13 students. Some of the girls had Marine boyfriends, who, of course were more important than their studies. I was drowning in a sea of TOO MUCH. Thank God Don's orders ending our tour finally came!

Looking back on that teaching stint, if I'd been there longer, I think I'd have surveyed the base for some parents who could do the subjects I couldn't and tried to get them to help. But Joan later told me she'd already tried that, without success. There were some correspondence courses available at the school, too, maybe from the U. of Nebraska. But no students were currently enrolled in any. Where there's a will there's usually a way, but I just had a bad case of "short-timeritis" and wanted to get that nightmare over with!

We finally got orders to leave in April, and actually, it was hard to leave Japan and our life of active duty in the Navy. Don would stay in the Reserve, which involved monthly meetings and a yearly duty stint at some naval air station, but we'd be going back to civilian life to make a living. We'd made some really good friends in the squadron and hated to say *sayonara*. Fortunately, we soon found out that no matter where VQ-1 people from that time were, we would still be friends and stick together through thick and thin. We still have that special bond today, even though most of us are in the 70's-80's.

Our trip back to the States was exciting in a couple of ways. The morning of our scheduled departure, in a last-minute check of our apartment, we found three drawers of unpacked clothes, all too good to leave!! We had to catch our plane, so we rooted through our garbage and resurrected three cardboard suitcases with "Lee Kee, Tailor, Hong Kong" printed in their sides and hastily stuffed those clothes into them. (These "bags" were leftovers from some clothes shopping we'd done in Hong Kong.)

We just made the flight, from Iwakuni to Tokyo, and then were to go to Hawaii via Midway and then on to CA. Well, as we traveled from Tokyo to Midway, Don looked at me and quietly said, "We just lost an engine." I was scared stiff, because we'd just seen a base movie about a flight that had had some bad mechanical problem beyond the point of no return and just had to suck it up and try to make their destination anyway. I can't remember the movie's name, but I think John Wayne was the pilot. Of course you know Hollywood' anything's possible, and the movie plane did make it back, after lots of nail biting in the audience. But to be sitting in a plane that the same thing was happening to nearly made me panic!

Sure enough soon a voice from the flight deck said, "We are now returning to Tokyo because of mechanical difficulties." And that was it! No muss. No fuss. We just turned around and went back. A couple of days later we got to set out again. We refueled on Midway, and on our next leg, I got sick and got my clothes dirty. So when we landed in Hawaii, Don turned them into the base laundry and we went to bed at the base. At 6:30 the next morning, there was a loud knock at our door. "Lieutenant Burdg?" a loud voice said.

"You and Mrs. Burdg are to report to the flight line at such and such a place at 0830."

"Well, Don quickly asked, "Are you going to send me my laundry I just turned in last night?"

There was silence for a bit, and then the voice came back to say, "We've taken you off today's manifest, but we'll send you out as soon as we have space."

Well, it turned out they didn't have space for a week! We had to stay at the base every night and check with them every morning, but for a week we were on Oahu with nothing we *had* to do! We knew families there and had some good visits with them as well as some very interesting tours of the island. All in all, that dirty laundry did us proud!

When we got the NAS Alameda, Don was released from active duty. Our next task was to job-hunt in Oregon, where we'd decided we wanted to live. Don got a job as a math instructor at a community college in Bend, central Oregon, but before we moved there, we needed to visit my dad in Colorado then Don's parents in North Dakota, and pick up a new car. Being short of cash, we went from Colorado to North Dakota by bus. I stress "BUS" because, with those Hong Kong cardboard suitcases and all the others we'd hastily packed in Japan and more in Colorado, we had *18 pieces* of luggage to check on the bus we took to ND! Amazingly, they didn't charge us extra, but Don was immediately out there helping our driver load all that stuff, so we could make it. We"d stepped back into civilian life.

Epilogue

August 1956: We'd tried to cram those 18 suitcases into a tiny apartment in Bend and were trying to adjust to living on community college instructor's pay. This was less than half of Don's Navy salary. And there was NO commissary for cheap groceries, NO Navy Exchange for good buys on appliances and hundreds of other items, and NO cheap Navy medical care. We had to live on a very tight budget!

Then suddenly one afternoon in late August we heard devastating news over our radio. A plane from VQ-1 had been shot down by the Chinese Communists over the South China Sea, with all members of the crew of 16 missing, although four bodies had been found, along with some of the plane's wreckage. We were stunned! No VQ-1 plane had ever been shot down until this happened. With mixed feelings we kept searching the newspapers for the names of those killed. We had to know, and yet we dreaded finding out just who had been lost. Our worst fears were confirmed when we read the list of crew members. Co-pilot Frank Flood, one of our dearest friends, was listed, and we knew a number of the rest as well, even though some had come to the squadron after we had left. Our hearts went out to all those wives and children of the lost aviators, as well as to all the rest of the VQ-1 personnel.

I was haunted then by the memory of a rather eerie happening on the previous New Year's Eve at our squadron party. I could see in my mind's eye one squadron officer taking out his cigarette lighter and acting as if he was going to set fire to some crepe paper streamers draped above the tables as decorations. It seemed such a strange thing to even think of doing. Those near him, including me, stopped celebrating and just watched him. While gesturing with his lighter for probably just a few seconds, he had this weird, dark smile on his face. I was close enough to see his dark eyes, and they had a strange glint to them. Then he put the lighter away and the party continued.

No one knew then that eight months later this man and 15 other crewmembers flying a night mission in a squadron P4M would be shot down by the Chinese communists, with all hands lost at sea. When we heard the news of this tragedy and read with awful dread the names of that crew, I remembered the look in that man's eyes and wondered whether he'd had a sudden premonition that New Year's Eve, 1955 would be his last.

We sent telegrams, letters, and our prayers. (This was before the days of satellite communications, so we couldn't phone or send e-mails). We were heartsick. In the backs of our minds, we squadron wives had always known an accident could happen. But we'd never talked about it. We had no idea what those secret missions were or how much danger our husbands were in every time they went out. Now here it was, an absolutely shattering loss.

In the years since then I've heard more and more details surrounding that terrible, yet still secret mission as the Navy has declassified some information and as Don and I have gathered for some wonderful reunions with squadron members and wives of that time who lived and worked in both the Philippines and Japan. Of course we've reminisced about the good, sometimes crazy times we had in 1954 to 1956. But those P4M missions, and especially that fatal one and its aftereffects have always come up and are relived again. Actually, time seems to warp and there we all are, back at Sangley Point and Iwakuni again.

Spring, 2001: We were in Thailand on a pleasure trip when the news reported that a communist Chinese jet fighter had crashed into a US Navy plane flying over the China Sea. The Chinese plane had broken into to pieces and fallen into the ocean. The US Navy plane severely damaged, had made an emergency landing on Hainan, an island off the coast of China." The crew had been taken prisoner, and lots of protests and counter protests were flying back and forth between our government and China's. As soon as we heard this, Don said, "I'll bet that was one of our squadron's planes." Sure enough, it was. Only now VQ-1 was/is based on Whidby Island in WA. And this incident had a much happier ending than the one in 1956. Diplomacy spared the captured crew's lives, and all were returned to the States safely. Thank God! Today, in 2009, VQ-1 is still based at Whidby Island. The full squadron name is now Fleet Reconnaissance Squadron One.

Spring 2006: The 50th anniversary year of the 1956 tragedy: Dean North, a thoughtful friend from VQ-1 in the 50's sent us a copy of The Washington Post Magazine. On the cover is a picture of Beverly Deane Shaver and her first husband, Jim Dean, taken at their wedding in 1956. The cover story, called "Truth and Lies," by Katherine Shaver, reveals that 21/2 months after their wedding, this bride and groom were in Japan, where he was a junior officer in VQ-1. He'd been on that ill-fated flight in 1956, and his bride and all the other crew wives were told that only four bodies of crewmembers had been found and that their husbands presumably did not survive the crash. Any questions Bev Deane and the other wives had about that crash were met with the same Navy and US Government response; "Sorry, classified information." A year later the wives were told that their husbands had been declared dead.

Bev Deane had gone back to her life before marriage and had become a pediatrician. She later married another physician and was having a good life with her new husband and their four children.

But in 1992, 36 years after the plane went down, Bev Deane Shaver, in a book store, quite by accident glanced at a new book called ***Soldiers of Misfortune: Washington's Secret Betrayal of American POW's in the Soviet Union***. Scanning the index under China, she found and read two paragraphs about a Cold War incident in 1956 off the China coast that ". . . nearly brought her to her knees." The VQ-1 incident was mentioned, and, according to these authors, two Americans had been captured by a Chinese patrol boat, taken to a Chinese hospital, and then to a

Chinese prison. The description of one tall, thin-lipped aviator sounded like Jim Deane. Bev was horrified to think that her first husband might have been tortured and imprisoned in China and never freed.

She was still very shaken when she told her second husband this shocking news. After talking it over with her, he urged her to go for a new search, which she did. This led her again to stonewall answers. Finally she contacted then US Secretary of Defense Donald Rumsfeld, a Navy flight-training friend of Jim Deane years before. Rumsfeld and his staff searched past records, but found no confirmation for the stories about two aviators being captured and then sent to prison.

Still determined to find the truth, Beverly then went to China. Her visit with a government official brought some stories which noted that many government officials from 1956 were dead, and the ones she was able to meet with kept changing or recanting their stories. They could find no records to confirm the men being taken prisoner.

Beverly Dean Shaver spent 13 years, from age 60 to 73, futilely trying to find out the truth. Finally, she was worn out and ready to quit searching, but one of her daughters, Katherine Shaver, a staff member of ***The Washington Post*** wanted to continue the search and did. She even found and interviewed some of the pilots in VQ-1 at the time of the shootdown. These interviews revealed some chilling facts about those mission flights that I'd never known. The missions were called "ferret flights" because the VQ-1 P4M's were actually trying to ferret out the Chinese radar locations by getting them to turn on their radar to track these dark silent intruders who flew over the ocean near their coast. With the Chinese radar on, electronic crew members aboard the VQ-1 planes could "ferret out" the locations of this radar and this Top Secret information was then sent to the very top echelon of our military in Washington, D.C. At times the Chinese would send out their fighter planes in attempt to chase down these dark planes and possibly shoot them down. But the large unmarked planes would make quick power dives to such a low altitude that the Chinese jets couldn't get under them to attack.

"What my mother didn't know, "reports Katherine Shaver, "was that her husband and his squadron were flying some of the most dangerous and secretive missions of the Cold War." I was stunned and chilled to learn these facts myself describing what these Shadow Warriors had actually done for years, losing 126 aviators in the process, from squadrons all over the world. I realized then that the fact that my husband and his crew had come home safely time after time was really just the luck of the draw and the good Lord's protection. These men had risked their lives, on every mission, to help the US win the Cold War then going on with China and Russia.

Katherine Shaver's article concludes with the fact that her mother, Beverly Deane Shaver, had finally come to accept what most others now believe to be the truth that Jim Deane did not survive the crash. But what a story!

July 2009—VQ-1 today. Several of us VQ-1 wives have written here the versions of what we remember about being attached to this squadron in the mid 50's. Commencing in 1992, one couple from this group, Stretch and Pauly Varner, very graciously invited us to their home in California for a reunion. Some 30 of us came. We had a great time laughing and talking, refreshing memories of everything from hilarious pranks, to the impressive military wedding uniting squadron pilot Glen Carlstrom and his bride Marilyn, to some of the adventures of flying and maintaining the P4M. Of course we also soberly reviewed memories of the shootdown and its aftereffects.

We wished that the wives of the lost crewmembers had come, too. We'd grieved with all of them and still were in touch with one or two. Of course we realized that getting together was probably too painful. And 36 years later they'd all gone on with their lives.

Well, the great success of that first reunion started a pattern, and nearly every other year since then we've had more, from coast to coast, and even a cruise. Of course "age related causes" have noticeably thinned our ranks, but that same fun and caring squadron spirit was still there in 2008, when seven of us got together in Oregon. Hopefully we'll get together again — to enjoy refreshing those memories and dealing with them, as we did then, with laughs, hugs, tears, and pride in what our men did to protect our country. And once again we wives know we'll feel that bond of friendship and courage that kept us going: living and thinking about what could happen, but never talking about it, and now realizing that what we didn't know until 2006 was a blessing.

Margaret (left) and Don Burdg

Margaret Burdg

Autobiography

Aug. 28, 2010

Margaret Burdg was born, raised, educated, and married in Colorado. She is a retired elementary and high school teacher. She taught 3rd and 5th grades, as well as high school at U. S. Navy bases in the Philippines and Japan. As a Navy wife, Margaret was able to live with Don at these bases. Later she taught English and Latin at Marshfield High School in Coos Bay, OR. Her husband, Don, was a U. S. Navy "Shadow Warrior" pilot based in the Philippines and Japan from 1954 to 1956. From 1956 – 1994 he taught mathematics at Community Colleges in Bend, OR, and later in Coos Bay, OR, where he and Margaret still live. They have two children and six grandchildren.

In retirement the Burdgs are enjoying traveling to visit family and friends, American States they've not yet seen, as well as many foreign countries.

Virginia (Jinny) Martin

SANGLEY SAGAS AND NIJIMURA NOTES:
A Concoction of Memories of a Navy Pilot's Wife In the Fabled Philippines and Japan

After having received his orders to COMNAVAIRPAC (Commander Naval Air Force Pacific) six months before expected, Lt. Barney Martin, USN, proposed marriage to me on Santa Rosa Island Beach, near Pensacola, Florida, in August of 1954.

I had been flying as an American Airlines stewardess based in Tulsa, Oklahoma, for four years when I met Barney. He was based in NAS (Naval Air Station) Pensacola as a US Navy pilot testing Twin Engine Beechcraft and had flown to Tulsa in May of that year to be Best Man in his brother's wedding. His brother's fiancee' was a friend of mine, and we met at the wedding. Barney and I dated by air, meeting in several cities throughout the country the entire summer, and fell in love. I even was invited to spend the Fourth of July at his mother's beautiful summer home on Lake Michigan, in North Muskegon, MI. Barney had been urging me to visit him in Pensacola, and when the American Airlines pilots struck at the end of July 1954, I took the opportunity to go there. When he proposed, we had been dating only three months. It was during this period that he received his surprising orders. He had been ordered to NAS (Naval Air Station) Alameda, California, for six weeks' training for further duty. It was highly classified, and I was told to ask no questions. At the time we didn't know what was to take place

after that or even if we would be separated. Having been reared a small town Ohio girl, the only thing I knew about the military was from World War II memories of my schoolmates joining the service to fight for their country and after the war coming home to civilian life. Almost all of them had been drafted and were enlisted personnel. It never occurred to me that people actually made a career of military life, or that I would be part of it.

In the brief time before our chosen wedding date, I returned to Tulsa to resign from American and pack the few belongings I had for the Navy to pick up. Some would go into storage and some would be for our future home. Then came the flurry of activity to choose a wedding dress, a going-away outfit, and find and outfit a matron of honor and a bridesmaid. Next, I had to drive to Ohio to pick up my mother and continue to Michigan for the wedding ceremony. My father would drive up later. Barney didn't get leave until the day before the wedding. That meant one more task for me was getting the marriage license, using a lawyer friend as proxy. (Barney returned the fee in front of everyone attending the rehearsal dinner.)

We were so captivated by Barney's mother's summer home, we decided we'd like to be married there, standing in front of the picture window overlooking the lake. It seemed the perfect choice since we wanted a small wedding with only family and a few close friends in attendance. We were to have an Episcopalian service. However, Barney had never been christened, and that was a requirement of my religion before the wedding could take place. He had been born in Barnes Hospital in St. Louis, MO, and so was always called Barney by his family. Since for nearly twenty-nine years all his records had been written "Barney (NMI) Martin," we decided that would be his christening name. The morning of our wedding, the christening was done by Dr. Walter Tunks, former rector of the Firestone Episcopal Church in Akron, OH, who would also officiate at the wedding. My parents were the christening witnesses in a little guest house belonging to one of Barney's mother's neighbors.

Barney chose his brother John as his best man, and his wife Leigh was to be my matron of honor. He also asked Jerry Smith, one of his bachelor aviator friends, to be an usher, and he wanted me to find a single girl to be my bridesmaid. All my friends were already married or engaged, so I asked Sarah Hayes, a fellow stewardess. Although I had flown with her, I knew her only slightly and never saw or heard from her again.

So on 2 October 1954, before a picture window overlooking Lake Michigan, I became Mrs. Barney Martin, a Navy wife. By that time Barney had been informed that he would be stationed at Sangley Point, a Naval Air Station located on a sand spit protruding into Manila Bay from Cavite City, Republic of the Philippines. There he would be assigned to Squadron VW-3 Det A, supposedly a detachment of a Guam-based early-warning weather squadron.

Our first night was spent in the Grand Haven Hotel in Grand Haven, MI. To Barney's surprise, his brother had forgotten to place his luggage in the car trunk, so in the morning we had to wait for a couple of friends to bring his clothing. We spent our honeymoon driving across country, stopping at his former prep school in Culver, Indiana; in my hometown, Medina, Ohio; his in Salem, Illinois; and his sister's and brother's home in Tulsa, Oklahoma. A party was thrown for us at every visit.

Barney had purchased a new white Lincoln Capri auto, which we named "Igloo," and the rest of our motor trip was spent with stops at the Grand Canyon; Santa Fe, NM; and Los Angeles and Port Hueneme, CA, the latter to visit Navy friends. Barney had bought me an American Indian dress in Santa Fe, and we stopped at a store to buy a headband and feather. When we arrived at our friends' door, I was dressed in the outfit and since I had lived in Oklahoma, they thought Barney had married an American Indian. Later in life I was to learn that I did indeed have American Indian heritage through a fourth great grandmother. I do not use the politically correct "Native American," because I feel that everyone born in America is a Native American. The first white men did refer to the natives as American Indians.

Lincoln Capri in white. Photo Consumer Guide.

After we reached San Francisco, we rented a furnished top-floor apartment in Oakland, CA. We lived there from the middle of October until the end of November while Barney was attending the highly classified school at Alameda. Again I was told that from that time on I was never to inquire as to what Barney was doing or his line of work, and I never did. Years later I was to learn that he was being indoctrinated into the world of Naval Intelligence Communications and receiving top-secret clearance. For me, those six weeks were spent preparing for our life in the Philippines (P.I.), buying appliances, and Christmas shopping for family and friends before we left. We made two mistakes. One was that we bought all Westinghouse appliances, and the other was that we purchased a Kerry Blue Terrier puppy in Hayward, CA. Barney had a friend with a Kerry Blue and had his mind made up that it was the breed he wanted. We drove there on Halloween and brought "Spook" into our lives. He was small and adorable at six weeks of age, and we smuggled him in and out of our apartment in a hatbox because animals were forbidden. That small, adorable puppy grew into a huge, inbred terror. Friends kept him in San Francisco until we were settled in Cavite and then flew him out to the P.I. on Pan Am.

The end of November we boarded a military prop plane at Travis Air Force Base and headed for Honolulu where we spent three nights of hell in the transient quarters of Hickam AFB. The wall of our room, if one could call it that, ended about one foot below the ceiling. In the room next to us was a family with a colicky baby who cried all night. Then a tidal wave hit with incessant rain as we were dining with a fellow naval aviator who had married one of my Delta Gamma sorority sisters from Miami University in Ohio. Their apartment flooded, and we spent the evening helping them bail out water. Not a good impression of Honolulu for a first time visitor, and since that time and several later visits, I have never cared for Hawaii.

Needless to say, I was delighted to board a military prop transport plane at Hickam to our final destination, Sangley Point. It was a very long, uncomfortable flight with only one stop in Guam to refuel. I don't recall how many hours it took us, but when we landed at Sangley my legs and ankles were so swollen it looked as if I had elephantiasis. This was very apparent because we were required to wear dresses when traveling on military airplanes.

Several Squadron members met us when we deplaned. After welcoming us, they announced that they were having a party at the Officers' Club that night, and we were expected to attend. I looked at Barney and rolled my eyes. We were both exhausted, but we went. One of the wives who met us was Ursula Gillooly, who eventually became a close friend. Ursula greeted me with the news that my husband would be flying in a plane as co-pilot with her husband Jack, another Naval Academy graduate, senior to Barney. She proceeded to say that it was a bad luck plane because it often had mechanical problems. Thank you, Ursula! I often wondered if that was the same P4M that was shot down in the China Sea after we left the Squadron. According to one article written about this Squadron by Bob Bublitz, "To Speak Of Many Things," there were nineteen P4M Mercator-type planes built by Martin for reconnaissance by the Squadron. (Ed. Note: Not all the planes were used by VQ-1. It used four or five of them; several were sent to VQ-2 in Africa, and some were never used at all. The "bad luck" plane ***was*** the one lost in August 1956.) Of course all the other Squadron wives had been told never to ask questions. From a distant roadway, I saw only the stark, dark blue planes with tail numbers and few markings. Bublitz wrote that the numbers on the planes were changed every month. There were guards around the Squadron area, and I seem to recall that the Operations building was a Quonset hut from WW II days, as were many of the quarters at Sangley.

We wanted a house in Cavite City that would be located just outside the Base. While hunting for one, Barney and I stayed at a lovely new hotel in Manila. I believe it was named the Bayview. The hotel was located near the pier where the boat carrying military personnel to and from Sangley Point was docked. Also close by were both the old Manila Hotel with bullet marks still left from WWII and the posh Army and Navy Club (see original coaster below). Later we joined the Army and Navy Club where we frequently enjoyed some leisure time.

Barney was only a Navy lieutenant and didn't have high enough rank for Base housing. There was a Filipino man who helped naval personnel find housing in Cavite. I am sure he was handsomely paid by the owners if he procured a tenant. Well, the first house he showed us had a kitchen with a dirt floor. I had a fit. I told Barney he could live in those conditions if he wanted, but I was staying in Manila at the hotel. However, after looking at several other properties, we decided on a fenced two-story home across the street from the Red Cross headquarters and the Government Building. Cavite was the capital of the province and that was what we would call the county court house. The house we rented was on a corner, and on the other corner was a building on stilts which was The Masonic Temple. There I first saw little boys running around in their shirts and in the buff beneath the shirts, playing under that building. That was to be just one of many shocks of Filipino living. The house located at #2 Rumaldo, Cavite City, was OK but it had no hot water and no screens. I figured we could boil our water, and taking a shower in cold water would be no problem in the tropics. The wall around the house was made from old runway matting and the yard was, as with most Filipino homes, just plain dirt with several fruit trees. I can still recall those Filipinas sweeping that blasted dirt in front of their nipa huts with those stick brooms.

Since we thought we would be there for a two-year tour (we should have known the Navy better), we would install screens, paint the house, and place a cement entrance. Things like that in the P.I. were still inexpensive in those days. The odd thing about the house was that it had a shower upstairs but no water. Barney had a flight to Hong Kong where he bought a pump, and we had a tower with a water tank on top installed to pump water to the second floor. By the afternoon, the water was warmly heated by the sun in order to shower. That was our first experience with solar heat. The mother of the house's owners used to think it was still her house and she would enter unannounced and pick the fruit from the trees. It

Drink coaster saved from the Army and Navy Club, Manila

amused me. I was told not to take my books out there but mistakenly did. There was a bookcase built into the stairwell where I put my treasured school yearbooks and others. I took one out one day and to my horror found it had been invaded by termites, as were many others. Fortunately we found them before all were totally destroyed. When I get one out to read since then, I am always reminded of our Cavite experiences.

Help was easy to find. We were fortunate to find a live-in housekeeper named Rose. She was Visayan from Leyte and spoke a different dialect from that in Cavite. We later learned that there were many dialects in the Philippines, and even the people in Cavite used words different from the ones in Manila. For instance, they were still using the Spanish word "mesa" for table. The national language was Tagalog, but not everyone spoke it. I tried to learn it, but the pronunciation was impossible for me, although I can still say "Merry Christmas" in the language. Later we hired Martin Ong, a house boy who was half Chinese and Filipino. He and Rose fell in love and later married, I was told. We asked Rose to return to the States with us, but she refused. After we were back in the States, she wrote a letter asking for money. Martin stole Barney's Rolex watch, so Barney wasn't interested in helping Rose. But she was a sweet and good girl. I always will remember her waxing our tiled floors with paste wax, and with her bare foot on a half coconut shell, moving it back and forth until the floors looked like glass. I wish I still had a Rose. To have a live-in maid was pretty nice for a newly married young woman.

After we were settled in the house with rented furniture, we sent for Spook and met his Pan Am flight in Manila. The flea situation in the tropics is bad, so we bought a big round galvanized tub at the local market and bathed Spook as often as possible. He was impossible to train. The only good thing Spook did was introduce me to Mrs. Gill, the wife of Sangley Point's Commanding Officer (CO), Captain Gill. Mrs. Gill owned a poodle and knew a groomer. She had the use of an official Navy car and driver and used to pick up Spook at our home to take him to and from the groomer in style. Good thing Congress didn't know about that!

Having been an American Airline stewardess for four years prepared me for Barney's flying. I never worried about him or his safety. I always knew in my heart he would return unharmed from his missions, and he did. Even in that unlucky airplane. I have always been philosophical about flying, but early on realized he was doing something "very" different and no questions were asked. However, at age twenty-seven I took his night departures in my stride and was always delighted when he returned, sometimes in twenty-four hours and occasionally two weeks. I always eagerly awaited his return from our separations. I never knew how dangerous his missions were until much later, after we returned to the States, when one of the planes was shot down with most of the crew with whom he had been flying all lost.

When Barney was flying, my first two months were spent writing thank-you notes for our wedding gifts. There had been no time in Oakland. Our friends and family had been most generous. There were so many, and I dreaded the job. In order to finish the chore, Barney had to bribe me with a beautiful white gold and diamond bracelet watch I had seen at the BX. I still cherish it. As a wind-up watch, it still works. Those were the days before battery watches. (BX = Base Exchange, also called Navy Exchange or Ship's Store; a sort of department store exclusively for military personnel and designated others.)

Before our Lincoln Capri arrived, we needed transportation to and from the Base. The Jeepnies, Army jeeps from the WW II era made into taxis, were not reliable enough; however we occasionally used them. It was the mode of transportation for Filipinos, with the driver charging a small fare. Barney purchased a Lambretta motor scooter for himself, and for me, a bicycle. I used that bicycle to go everywhere and carried loads of groceries to and from the Base Commissary in the basket. I still laugh when I think of our going to cocktail parties on the Base riding that Lambretta scooter. He was in his dress whites and I was in a cocktail dress and heels riding behind him, dodging Cavite, chickens, pigs, children, and prostitutes lining the street in front of the Base gate. When we left the P.I., we sent the motor scooter to Ohio for my father to use in our small town of Medina. He loved it and was quite the sensation riding it to and from work and around the small town.

The people in the Squadron were all super and made us feel quite welcome. Most of them had small children, so Barney and I didn't have much in common with them in that department. We had decided not to have children while in the Philippines, due to lack of good medical care, so took precautions. Little did we know that was quite unnecessary. John Ebersbaker was one of the electronics officers in the Squadron, and we became close to him and his wife Helen, who were older and more able to travel. She had grown children in the States by a former marriage, and John was several years younger than she. We traveled with them to Baguio, the P.I.'s mountain resort, and to Hong Kong. In Baguio we stayed at a military R & R (Rest and Recreation) complex named Camp John Hay and played golf. I always recall one hole that was called "cardiac hill." It was a perpendicular hill below a green which was impossible to penetrate. I always took the stroke on that one. In the distance we could hear artillery firing by the Huks. Their full name was Hukbalahap, and they were Communists fighting the government to take control. They never succeeded, but the Philippines always appear to have some kind of guerrilla organization battling the government for control.

I soon became close friends with Irma Kruzich, one of the Squadron pilot's wives. She was quite adventurous, as I am. She was talented and loved to sew; we both loved to shop and had the same tastes. One day Irma told me she knew that we could get some beautiful big green jars to make into lamps at the local Filipino Military Station in Cavite, and that they had lots of them. So we toddled over there and introduced ourselves to the CO. When Irma said her name was Kruzich, the Filipino thought she had said Cruzen, so he was all smiles and told us to help ourselves. We chose two bottles each, which was all we could carry. Admiral Cruzen was COMNAVPHIL/CINCPACREPPHIL (translation: COMMANDER NAVAL FORCES PHILIPPINES / COMMANDER IN CHIEF PACIFIC / REPRESENTA-

TIVE PHILIPPINES), a "BIG" wheel, and lived in beautiful large white quarters located on the Base overlooking Manila Bay. He was very generous about entertaining the officers and their wives stationed on the Base. He also lent the Squadron his sixty-five foot yacht **Margaret**, operated by Special Services Dept., NAS Sangley Point, for cruises to Bataan and Corregidor. Apparently Admiral Cruzen was notified of "his wife's" visit to the P.I. Base, and all hell broke loose. Our husbands were chastised, but we did keep the bottles, which I still have.

The day our Lincoln Capri arrived was so exciting. As we watched it being unloaded from the ship in the Manila harbor, our excitement turned to horror as we saw them drop a huge piece of metal on the hood, crushing it. Our motor scooter days were longer than we had anticipated. After it was repaired, we sold the car to a wealthy Filipino doctor. I always wondered what happened to it, for few Capris were manufactured. It would be a collector's car today. But we soon learned that the Filipinos had three generals. They were General Mc Arthur, General Motors, and General Electric. So the stupid Martins arrived with Westinghouse appliances, a Ford motorcar, and we were in the US Navy. Not good planning for reselling.

One day Irma informed me that she would be teaching a Red Cross women's adult beginners' swimming class. She had been reared in Panama and was a former Olympic swimmer. I was terrified of water because I had a bad experience at age ten. A YWCA instructress pushed me into the pool when I wouldn't jump. She then pushed a ragged bamboo pole into the water for me to grab because I was drowning, and it cut me on the thigh. She finally had to jump in, pull me out, and pump the water from my lungs. That had been the last of my swimming efforts. But Irma could be a very convincing person. Not only did I learn to swim in her class, I completed learning the skills in the Intermediate and Swimmer classes taught by Ursula. Pauly then taught the Senior Life Saving class. Within six months I was ready for the American Red Cross Water Safety Instructor Course

Warm-up exercises before swim class

taught by Pauly. She also had a water ballet class which was a lot of fun. I owe them all a debt of gratitude. Not only did I become a Red Cross swimming instructress myself, but the long days spent at the pool helped me cope with being a bride with only a part-time husband. One interesting aside was the fact that both officers' and enlisted wives could take part in the courses. I met a lovely wife of one of the Squadron's enlisted men, and she invited me to their quarters. I accepted the invitation, thinking nothing of it. Almost all of my friends in World War II were enlisted. When my husband returned and I told him about my experience, it was then he admonished me, explaining that naval officers and enlisted personnel and their families did not socialize. It was one of my first Navy protocol lessons, the first of many lessons learned on how to be a naval officer's wife.

All the parties the Squadron gave were held at the Sangley Point Officers', aka "O", Club. The "O" Club was a converted Quonset hut and a lively place. The party I most remember was the Klondike Gold one. Everyone came in costume worn in that Gold Rush era. I had a gold lame' dress made from some material purchased in Hong Kong and went as a gold nugget. A play was given by Squadron members based on "The Cremation of Sam McGee." It was hilarious and, of course, rather alcohol-filled with laughter. Naval aviators had to let off a little steam after their dangerous missions.

The other occasion there, which we probably all remember, was the cocktail party given for the then Chairman of the Joint Chiefs, Admiral Arthur W. Radford. Admiral Radford was instrumental in building Naval Station Subic Bay at Cubi Point, west of Manila on the northern side of the Bataan Peninsula. He was in Manila to meet with President Magsaysay at his Malacanyan Palace. The night before the meeting, Sangley Point went all out to entertain him with one of the most sumptuous buffet dinners I have ever seen. To us lowly junior officers and wives, this man was God. We turned out in our finest and stood in line to shake his hand. Barney and I had been invited to a dinner in Manila, so we met the Admiral, had one drink, and left without eating. The next day, the Squadron had planned a cruise by Fort Drum and on to Corregidor on the yacht *Margaret*. Seeing these places and having known of the horrors there in WW II was always an emotional experience. This was ten years after the war ended and Army military relics were there, overgrown with native plants, to remind us of the last stand and the terrible Bataan Death March. The day of the cruise, suddenly one by one, almost everyone started to become ill, vomiting over the vessel's side. At first we thought they were seasick, but then they started having problems at the other end. Only a few of us, including Barney and me, escaped Montezuma's Revenge. The ones hit were those who had eaten the food at the party the night before. Later we learned that even Admiral Radford had been stricken and found it necessary to cut very short his meeting with President Magsaysay. So it was food poisoning of some kind. This was not surprising with such lack of refrigeration facilities at the time and the amount of food that was prepared in advance and eaten.

Admiral Arthur W. Radford was in the Philippines to check on his project of creating a modern operating base with an airfield and anchorage for deep-draft naval ships on the Bataan Peninsula of the island of Luzon. It would be on land permanently ceded to the United States in its former colony. The Navy's Seabees took on the project, and in five years of round-the-clock operations, Construction Unit Battalion 1, "CUB1," completed the job. It involved twenty million man-hours and flattening a 1,200 foot mountain to construct an 8,000 foot runway. Some say the level of effort was the equivalent of building the Panama Canal. In July 1956, Admiral Radford returned to Cubi Point, near Olongapo, to commission the new station. We once drove up there to see the progress which was well on its way for completion while we lived in the Philippines. Ships at that point were able to dock there, and we took one out of that port on an R & R for our first trip to Hong Kong. The Naval Air Station has now changed its name to "Arthur W. Radford Field."

We often drove to Clark AFB which was located on the way to Baguio. This was like going to a country club for us. It was obvious which service was getting the taxpayers' money in those days. The quarters, Officers' Club, and BX were magnificent compared with those at Sangley Point.

Our first Christmas tree in our home at #2 Rumaldo was pretty sad, but at least it was a tree. Those were the days before fake trees. This poor thing had spindly branches with few needles and decorations, but at least we had a tree for our first Christmas together. Fortunately, I have a photo of it.

When we first arrived in Manila we hired a driver to show us the sights. I shall never forget seeing Santo Tomas, where WW II prisoners were jailed, tortured, and sometimes killed by the Japanese. The driver pointed to a beautiful red hibiscus bush and said that the people ate the flowers during the war because they were so hungry. I think of that every time I see a hibiscus blossom. During our Sangley Point tour, we visited Manila often and toured Malacanyan, the president's palace. Barney's Aunt Bertha, his father's sister, had been married to a lumber baron and lived in the P.I. before the war. She told of having dinner at the palace one night, and as she was about to eat her soup, a gecko's tail fell from the ceiling directly into it. Another place we visited was the San Miguel Beer Distillery. It was good beer. One nice thing that happened to us was receiving an invitation to lunch at the American Embassy with the Ambassador Homer S. Ferguson and his wife. He was a former Michigan senator and friend of my mother-in-law's neighbors in North Muskegon. They arranged the invitation, which was quite a feather in the cap of a young Navy lieutenant and his wife. The Fergusons were charming, and I will always recall one of his comments. He said the Filipinos had chosen Americans' bad habits, but not their good ones. For example, he said ten years after the war Manila had no public library. However, the Filipinos had many fine qualities and later re-established their educational system.

Irma Kruzich was always coming up with interesting things to do. We loved shopping in Manila. We were both artistic. One of the many talents of the Filipinos was wood sculpture and carving. They are masters in that art. Irma had met a talented Filipino wood sculptor named Joe. If Joe had a last name, I don't remember it. We both started to take lessons from him and used a piece of Nara wood mahogany for a mallet. I purchased some inexpensive chisels. I still have the unfinished cat I was sculpting. I also purchased one of Joe's beautiful pieces which I still cherish, the bust of a Mora woman. I learned a lot about the different kinds of mahogany wood and also ironwood which many of the carvers in Baguio used.

Another Filipino talent includes embroidering. I became fascinated with the Filipino men's dress shirts called Barong Tagalogs. Some were made of fine cotton and some of pina cloth. Pina cloth was a fine material hand-woven from the fibers in the leaves of the pineapple plant. It was light beige in color and had a filmy appearance. Barney and some of the other officers had them made and wore them to casual events where uniforms were not required. They were a pullover shirt with a tailored collar and several buttons from the collar downward about a foot in length. Beautiful hand embroidery was around the button seams as well as on the cuffs. I had one made for myself, along with a dress with a barong type blouse and a full skirt for cocktail wear. I also had a Barong Tagalog made for myself from a soft white cotton material. I still have it and wear it occasionally. They are difficult and time-consuming to iron.

The Filipinas are also beautiful seamstresses and were good at smocking. Many of us found seamstresses in Cavite and had lovely smocked summer dresses created. Unfortunately that type of dress later went out of style, and I gave them all away. I still have a traditional Filipino formal gown that a Filipina seamstress created from material I purchased at a local store. Their native women's formal costume is a beautiful floor-length gown with large puffed sleeves and is heavily embroidered. I sometimes wore it later in Washington, DC, to Filipino Embassy parties. I believe it was called a Baro't Saya or Maria Clara.

I made friends with a young American woman named Marilyn Lippard whose parents lived in Manila. Her stepfather was a government engineer there to oversee a TB hospital being built in Manila and assure the US government it was being built to proper specifications. Marilyn met and dated one of our bachelor officers, Glen Carlstrom, and when they decided to marry, I was flattered that Marilyn asked me to be her matron of honor. It was a beautiful wedding in the Sangley Point Base chapel. After we moved to Japan, they were our neighbors and we saw them often.

One weekend Andy and Irma Kruzich invited us to go with them to Pangasinan Falls. Barney and Andy had arranged the trip, and I only knew it would be some kind of a boating venture. As we were dressing, Barney mentioned what I should wear since we were going to be shooting "rabbits." I replied that I wasn't going to go then because I didn't want to shoot those cute little bunnies. He roared with laughter. He had said we were going to be shooting rapids, and he never let me forget that one. We boarded a small boat on a mountain stream with many rapids, and every time we went over one Barney would point his finger at it, shout "Bang, bang," and laugh hysterically.

Another tourist attraction we visited was Lake Taal and the beautiful Hotel Tagaytay. I have a moving photo Barney took of me walking down the steps of the hotel toward the lake. From the steps of the hotel one can see that there is an island within the lake. I am told there is also another lake within that island, but it can't be seen by land. We also witnessed a performance of Filipino pole dancing there. This is an athletic feat of dancing as low as the dancer can go while the poles held by two people on the sides are slowly lowered to the floor until they can barely dance beneath it. They actually are dancing with their bodies parallel to the floor.

Occasionally we were given R & R trips in ships to Hong Kong. Barney and I were fortunate to have been able to travel on one with the Ebersbakers. We stayed at a hotel named Miramar in Kowloon. It is no longer there, but at the time was one of the most modern. There were Indian tailor shops in the lower floor, and we purchased material and had clothing made by one of them with whom we became friends. We could take photos from fashion magazines to the tailors who would copy them perfectly. We also had shoes made and visited the island of Hong Kong, Tiger Balm Gardens, and other tourist spots.

Some of the Squadron members had found Mr. Lee, a furniture manufacturer in Hong Kong who did beautifully hand carved non-lacquered teak furniture and rosewood silver chests. Silver doesn't tarnish in rosewood. Since we were newlyweds, we had no furniture, so with my BFA (Bachelor of Fine Arts) background I designed a set of bedroom furniture and showed Mr. Lee the plans. He said he could do it with no problem. There was a hand carved king headboard in two pieces with sliding doors. To place small items, he made a makeup table with a teak mirror frame, and there were three Mr. and Mrs. chests with Chinese brass hinges and locks. I specified that the two chests on each end should be engraved with Chinese characters, one saying "Mrs." and the other "Mr." I wanted everything done in parts so they would be easier to move. I had taken all of my wedding silver with us to the P.I. and left it in Hong Kong, with much trepidation, under Mr. Lee's care. He was to make a rosewood silver chest with spaces handmade for each piece. Until I picked it up on my next visit, I was worried sick that I would probably never see that silver again. I was wrong. Not only did Mr. Lee create a beautiful bedroom suite for us that has withstood time and moving, but the silver chest was just as he said it would be. I found another one the same size to match, but without the specially made places for each piece. When I returned to pick up the furniture, Mr. Lee and his lovely wife, both speaking perfect and fluent English, treated me to lunch on the Hong Kong Island, and I learned much from them about China. Mrs. Lee told me that there were so many dialects in China that when a cousin visited her in Hong Kong from another province, they had to communicate in English because they could not in one another's Chinese. The furniture arrived safely in the P.I. It now dwells in the home Barney and I built in Rancho Santa Fe, CA, after he retired from the Navy. The bedroom was designed to fit the furniture.

I also purchased many other treasures in Hong Kong, such as two small jade-decorated jewelry chests, teak nest tables, and other small pieces of teak furniture, as well as about eighty abaci boards, the Chinese adding machines of the day.

These were later made into a screen and used as legs for a coffee table I designed. I also found several Chinese lamps, two made of abaci boards, and several of porcelain. Ivory was available, and regrettably I purchased some knick-knacks and jewelry made of it. That was before I realized the price paid by an elephant to furnish it. All of it has now been sold.

Barney was unable to get leave to return to Hong Kong with me, so I went with Helen Ebersbaker. Our little Indian tailor friend again made us some outfits, and we bought a lot of material from him to take home. He invited us to have dinner in the apartment where he lived with his brother (or cousin) and wife. When we arrived, there was a little coffee table set with a bottle of Seagram Seven Crown situated in the center. We sat on the floor around the table, and the food that was served was so spicy that I could barely swallow it. Maybe that is why the whiskey was there to chase it down. I had bought too much material for my suitcase so I carried it on my lap on the plane, telling the officials that they were my lap robes, which were allowed to be carried. It was a British plane flying between Hong Kong and Manila.

When we first arrived in Sangley Point, the Commanding Officer of the Squadron was a man named "Hap" Sweeten, and his wife's name was Pink. We barely knew them since they left shortly after we arrived. Jack Reynolds became the CO after Hap, and when he and DaLoris left, an officer named Gene Hall arrived to take his place. Gene came ahead of his family to see that things would be prepared for them, so Barney and I thought it would be nice to invite him to our home in Cavite for dinner. Barney told me that Gene drank gin martinis and that he would have to go to the Base to buy some gin. I had seen a bottle of gin in the refrigerator and told him that we already had enough. After Gene arrived and was seated, Barney took the gin bottle from the fridge and made him a martini in a gorgeous glass with olive, "et al." Gene took one sip, and this strange look came over his face. Well, our Filipina housekeeper Rose had taken a gin bottle, filled it with water for her use, and put it in the refrigerator. So there were no martinis that night. Perhaps that incident contributed to Barney's poor fitness report when he left the Squadron.

A dark side of living in the Philippines was hearing the moaning and crying of dogs at night. We were told that the Filipinos stuffed them with rice, beat them to death, and roasted and ate them. Also, for some reason, they preferred black dogs. Spook was black, and one night he disappeared from our gated and fenced yard. Barney immediately went looking for him and had a temper. He found two Filipinos lurking around the house and told them to get the word out that if our dog wasn't returned unharmed, he would kill them both. The men were terrified, and Spook was soon returned to our yard.

We lived not too far from the Cavite market, and I used to walk there to look at what was for sale. However, I found it disagreeable and was surprised to see so many Sangley Point BX items that had been resold to the merchants for Black Market items. I preferred to shop at our little Base Commissary, and even then we were told to wash our fruits and vegetables in a solution of water and Clorox, which I did religiously. I did

Native housing on lagoon in Cavite

my own cooking, and Rose served the meals. Except for lumpia, a Filipino appetizer consisting of meat in a small pastry roll, I really didn't care for Filipino food. Barney loved blueberry pancakes, and that became a Sunday tradition when he was home. That is, until the flight surgeon told him he had to lose weight or be taken off of flight status. I never made them again. My mother made wonderful pies and taught me how to make a perfect pie crust. I recall I made my first pie for Barney in our Cavite home, and when it was served he ate the filling and left the crust. I asked, "What are you doing?" and he said that he didn't like pie crust. I never made a pie again either. Such a waste of talent! (Smiley Face)

When we drove the Lambretta to Manila by land, I used to marvel at how beautiful the Capiz shell and mahogany windows looked with the lights shining through them from the nipa huts along the way. The lighting was enhanced by one long fluorescent bulb attached to the middle peak of the ceiling of most huts. All the shells were cut into small squares surrounded by the mahogany. They were used as sliding windows in the huts and were about two feet square. The artist in me said that if I had some of those windows, they might someday be of good use. Luckily I found a place along the road that made them by hand, including the mahogany frames. I bought about seventy-five of those windows for fifty cents per square. They later were used in a variety of ways in three of our stateside homes, removed and replaced as we moved, and presently are the main theme for our Rancho Santa Fe's oriental bed and bath suite.

Around the corner from our Cavite home was a house where the local prostitutes used to line up for their physical health checkups. It was called The "Clinica." On the way to the Base, I recall passing a horrible body of water with nipa huts on stilts hanging over the water. I was told that there were holes in the floors of the nipa huts, and that is how they did their toilet duties. Ugh!

Why Barney decided it would be a good idea to take Spanish lessons, I will never know. Nevertheless, he found a lovely Filipina teacher who came to our home. Some Filipino upper class people in Manila still spoke Spanish, but we certainly didn't know them. Barney had taken the language in the US Naval Academy and later in the Naval Intelligence School. He was miffed that he was sent from there to Pensacola to test Twin Engine Beechcraft, instead of to a Spanish-speaking country. I don't recall the name of the teacher, but she was a widow and told us of her late husband's capture, torture, and death by the Japanese. She said that before WW II there were many Japanese shops in Cavite, and the day the war broke out all of the Japanese came out of their shops and homes in military uniforms. They had all been spies. On December ninth, two days after Pearl Harbor,

Cavite burning after Japanese attack. Photo USN.

Japan declared war on the Philippines and bombed Sangley Point, destroying the Base.

I did enjoy looking in the various Cavite shops and one in particular which sold handmade wooden products. I **LOVE** wood. I know not why Barney and I were selected to choose going-away gifts for the departing officers and their families, but it was a challenge. I had seen some round Nara wooden trays in the wood shop and came up with the idea to have the name of the Squadron, the dates served in it, and the names of the husband and wife hand-carved around the edge. They turned out well, and everyone seemed to be pleased. However, that is when I learned that DaLoris Reynolds did not spell her name Dolores, so another tray had to be made. I still have and occasionally use ours, and I hope the other Squadron members have done the same. It is a lovely souvenir from an exciting era.

One day the aircraft carrier USS *Kearsarge* arrived in Manila Bay and anchored amongst the rusted hulks and masts of the Japanese ships sunken by the American forces when they returned to the P.I. during WWII. In the Pensacola BOQ (Bachelor Officers' Quarters), Barney had become friendly with a Catholic Priest named Father O'Looney. Barney discovered that he was serving as a Navy Chaplain on the Kearsarge and invited him to our home for dinner. I never knew his first name because we always called him Father. One vivid thing I recall his telling us was that when the carrier anchored in Cavite, the VD rate among the men rose 75 %. After that I always questioned the nature of the physicals performed at the "Clinica." Father O' Looney reciprocated our dinner invitation with one aboard the Kearsarge. Now, this was a pretty exciting thing for a girl reared in Medina, OH! We had to take a boat out to her from the Sangley pier, and of course I was dressed to the nines with high heels and a cocktail dress. I had to be hoisted by the men up the "ladder," as stairs are called on a ship, by a couple of men. Then I hobbled aboard, clinging to my fluffy skirt. That was a real "WOW" for me. He gave us a tour of the ship, and the communications room was the most fascinating for me. Since then I have visited many carriers and the marvelous Midway Museum in San Diego, but that first time will live in my memory forever. I often wonder what happened to Father O'Looney, for we never saw or heard from him again.

It was quite exciting to learn that VW-3 Det A would be commissioned as a full-fledged squadron named VQ-1 in June 1955. A new squadron needed a logo, so the CO announced a contest would be held amongst the members of the Squadron and their wives. Designs were to be submitted by a certain time. Being an art major, of course I had to get into the act. But what would I design? I had no idea of what these planes and pilots were doing. So I submitted a globe with a cloud background over water, islands, a ship, and three monkeys: see no evil, hear no evil, and speak no evil. The monkeys were supposed to represent classified material. The late Frank Flood, Barney's co-pilot, entered a design with a bat and electronic symbols. He had the edge on me because he knew what he was doing. The judges decided they liked both designs and put them together. They used Frank's bat and lightning on my background. We each received an award of a pen and pencil set as a prize. I have heard that the present Squadron has changed the design, which is a shame. It would be a tribute to Frank Flood to have kept it in his honor. He was in the ill-fated P4M shot down by the Chinese and lost in August of 1956. Those brave men, in my opinion, should have been awarded recognition and posthumous medals they so deserved but never received.

Original VQ 1 Squadron Insignia

The commissioning of the new VQ-1 Squadron was held on June first. The wives were invited to go into the forbidden territory where the planes were parked for the ceremony. I will always remember them in their whites, standing at attention for that big day. They each received a beautiful large Plank Owner's certificate (see below). I recently donated Barney's to The Naval Historical Museum at The Navy Yard in Washington, DC.

One afternoon Barney came home and announced that he had a big decision to make. Unbeknownst to me, he had applied for a designation change from aviation to intelligence duty. That is 1300 to 1630 in Navy terms, in other words a change from flying status with flight pay (1300) to non-flying status without flight pay (1630). After he left Intelligence School, he decided he would enjoy that instead

of flying, but his application was turned down. He thought that was the end of that and forgot it completely. Word was sent to CO Gene Hall that Barney had been picked up and chosen for a 1630 designator. Gene accused Barney of going behind his back to do it, and of course that wasn't the case at all. Barney also had neglected to tell me what he had formerly done. I told him that I knew too little about the Navy to advise him, and he would have to make the decision without my advice. I had flown in the airlines with many former military pilots and never thought Barney was the pilot type. In fact, I always said that I would never marry a pilot, but that changed once I became acquainted with Barney. So in September 1955, Barney made the decision to leave naval aviation and before we left Sangley Point was taken off flight pay and piloting P4Ms. Later in our marriage, I mentioned to Barney that it was a good thing he chose to become a 1630 due to the fact that had he stayed in the Squadron his second year as a pilot, he would be dead. He answered, "No I wouldn't, because I wouldn't have been flying where Hutch was flying". Hutch was the nickname for Lt. Cdr. Milton Hutchinson, the commander of the ill-fated plane shot down by the Chinese in August 1956. The incident was never mentioned again, and I never questioned him further. I did discuss it later with one of the pilots at one of our reunions, and he opined that the radar was so poor that "Hutch" probably didn't know how close he was to the Chinese coast. Unfortunately, this is something none of us will ever know.

What a surprise when it was announced that our supposedly two-year Philippine tour would be cut to ten-and-a-half months! Now that it was an official Squadron, VQ-1 — its planes, entire personnel, lock stock, barrel, and baby buggy — would be moved to Iwakuni, Japan. It was pretty exciting, but with one catch. There wasn't enough Base housing at NAS Iwakuni for about twenty of the Squadron's junior officers and their families - just one of many boo-boos in government planning. But there was obviously a good reason for the change.

My husband, dog, and I would probably be in Japan a short time because Barney would be doing desk work for the Squadron and soon receiving orders to leave, due to his designator change. So we scurried around, trying to make plans as to what to send home to storage and what to take with us to Japan, which turned out to be very little.

Our future was in limbo. I had purchased those abaci boards in Hong Kong and before we left Sangley Point I found a furniture maker in Manila to make us a three-paneled black lacquered screen of abaci. He also made a black coffee table I designed with two small tables under each end using abaci for legs. The little tables could be pulled out to make the coffee table larger. I had also purchased abaci lamps to match and had shades made of grass cloth on which I painted red Chinese characters. It was a must to take back home a set of rattan furniture. So I designed a three-piece sectional with a chair and stool to match and a two-tier end table of rattan with Nara wood mahogany tops. We also needed dining room furniture — remember we were newlyweds and had nothing — so I designed a mahogany buffet with three caned doors. To save room when not in use and look like it is part of the buffet, a mahogany table on large brass casters rolls over the buffet. The tabletop has hidden hinges so it can fold in the middle. Then, when rolled out from the buffet, it opens at the hinges, thus sliding into a middle, locked position on the base and four legs. This becomes a full table seating six people. I also bought a small mahogany and rattan table with chairs to match. All of this, along with our Hong Kong teak bedroom set, was sent to storage in the States for a future first home. Our landlord bought our refrigerator and freezer, which was convenient. We also ordered two automobiles to be delivered to us in the States: one, a Buick sedan to San Francisco, and a little British Morris Minor station wagon for Spook and me, to the Washington, DC, area. There Barney would be stationed in the Pentagon, learning the ropes of a Naval Intelligence Specialist. The Buick was a disappointment, for it was not at all the color we had ordered. The Morris Minor traveled all around the world before it would reach us in Alexandria, VA.

The men ferried the P4M planes from the Philippines to Japan, and Spook in his crate went with Barney. I don't recall how I was sent, but probably commercial air. When we arrived, we discovered that the families of the junior officers would live in Nijimura housing, a two-hour boat ride across the Inland Sea of Japan to the Base in Iwakuni. The men made the two-hour boat ride morning and evening, Monday through Friday and a half-day on Saturday. This meant they spent twenty-four hours a week commuting on that boat. The Navy Department didn't plan very well. I still cringe to think of what it cost the taxpayers.

It wasn't much fun for our husbands to have to do all of that boat traveling after a grueling day at work. Barney devised a plan where they would have cocktails during the trip on the way to Nijimura so it wouldn't seem so long. Apparently someone didn't like the idea and blew the whistle on them, so one day the little cocktail parties came to a screeching halt.

Nijimura was near a small Army munitions base in Hiro, Japan. The housing had originally been built by the Australians for barracks after WW II and somehow was taken over by the US Army. The Navy arranged for the officers and their families to be placed in those quarters. One of the problems with the housing was its location across a body of water from a pulp mill. The stench was terrible, but we got used to it. Each of the barracks had been converted into four two-story apartments. Glen and Marilyn Carlstrom lived in an end apartment in the building across from ours. We were fortunate to acquire an end unit, too. All were sparsely furnished and had no window coverings. I wished to make our apartment as pleasant as possible. Barney had bought me a sewing machine that I wasn't too crazy about operating unless it was something simple. So since we didn't know how long we would be there, I bought some inexpensive muslin at the Iwakuni BX and made some curtains.

The wives could ride on the boat to Iwakuni when they wished to shop on the Base or secure medical and dental treatment on the Iwakuni base. Dependents abroad were able to have military dental care. This was not available to them in the States. The Inland Sea boat ride was beautiful. The scenery was much like lovely, misty Japanese silk paintings. We purchased one before we left as a reminder.

There wasn't much of a social life at Nijimura, due to the long boating hours. The men arrived home to have dinner and then to bed in order to arise early the next morning, about 4:00 A.M. But usually something was planned on a Saturday night so we could sleep in on Sundays. There was a small Army Officers' Club at Hiro, and Barney and I occasionally ate dinner there on weekends. They allowed some Japanese sales people to sell their wares in the Club, and we purchased a beautiful hand sewn portrait of a Japanese man. The thread work on that piece of art is amazing. There was a matching female portrait and at dinner we discussed buying it, but when we returned it had already been sold.

Since I had become used to having a housekeeper, we hired a Japanese woman to come in to clean. I don't recall her name, but Spook did not like her. She giggled a lot and could not speak English. Well, she did something to displease Spook, and when he bit her, she started to giggle. I said to her I was glad she thought it was humorous because I didn't. I later discovered that when Japanese women were nervous or upset they giggled. I felt bad for the girl and was horrified that we would be sued, but she resigned and we never heard from her again. I then found another lovely Japanese girl named Suzi-San, who spoke and understood English. She had worked for other Americans and got along well with Spook. She cleaned and ironed for me until we left Japan. There was no wash-and-wear in those days, so the men's uniforms had to be washed and ironed. My household help in both countries did a beautiful job in that department. It was — and still is —not one of my favorite things to do, but after our return to the States I did plenty of it, washing and stretching those cotton uniform cap covers onto an adjustable metal frame to dry.

When he was still flying out of Sangley, Barney had bought me a beautiful Japanese kimono. He always was bringing me gifts from wherever he had been flying. From those gifts, it was no secret to me where he had been going. The kimono was a beautiful flowered pink and blue, but had no sash, or obi as the Japanese called it. So I decided to ride my bicycle, on the left side of the road that I sometimes forgot, to the little town of Hiro to buy an obi. Of course I spoke only a few words of Japanese but figured I could get by through pointing. I found the shop and pointed to a beautiful pink obi. The shop owner shook his head no. I showed him some yen. He still shook his head no. So I rode my bicycle back to Nijimura sans pink obi. I told Suzi-San about my experience, and she went back to the shop with me to inquire in Japanese what was wrong. The proprietor said that color was too young for me and he wouldn't sell it to me. It appeared that children in Japan wore the brightest obi and kimono colors possible, and as one aged the obi colors grew darker. I probably would be wearing black today. Suzi-San explained the situation to him, and he reluctantly sold me the beautiful pink obi.

We lived an hour's train ride from Hiroshima and decided to go see it one Sunday. The train system in Japan was, and still is, marvelous. It had only been ten years since the war had ended, and there was still some rubble there. They had already built a museum with photos and evidence of what the terrible Americans had done to them to save not only the lives of Americans but also Japanese. If it had been necessary to invade Japan, thousands more lives would have been lost on both sides. We visited the museum and of course were the only Round-Eyes in the place. The Japanese people kept staring at us, and I told Barney that I didn't care; it was their stupid government who started it all by bombing Pearl Harbor.

We had arrived in Japan in October, and Barney received orders to report to Naval Intelligence at the Pentagon in January. Since he was no longer flying, he asked for two weeks' leave over Thanksgiving, and we took the train to visit Nikko, Tokyo, Nara, and Kyoto. What a marvelous adventure! We took the baths in Nikko, and I still have a souvenir cotton kimono and sash we wore in the hotel. In Tokyo we stayed at the beautiful Imperial Hotel, designed by Frank Lloyd Wright. Unfortunately it has been destroyed, and I feel fortunate to have had that experience. Even then, Tokyo was a large, traffic-laden city. Later, in Nara, I marveled at the adorable deer in the temple parks. We spent Thanksgiving, which of course meant nothing to the Japanese, dining in a Japanese inn's private room, sitting on cushions on the floor with our cold feet heated by a hibachi, and eating a traditional Japanese dinner with dessert as the first course. I loved visiting the temples and historical parts of Japan and would never have the opportunity to go there again.

A Saturday night not long before we left Japan, the Squadron men living at Nijimura planned a dinner party in a local Japanese restaurant to show us wives how much fun they had when their missions held over at various places in Japan, and also a going-away party for those of us leaving for home. It was the traditional pillowed floor seating, and the dinner was being served by beautiful Japanese women attired in colorful kimonos. They undoubtedly were what was called "after (après) war geishas" — Japanese girls who were paid to serve and entertain but had not had the extensive training given a true geisha. As I was dining, I suddenly felt a hand stroking the back of my hair. Back then I had very shiny, fine, brunette hair. I looked around, and it was one of the waitresses. She said "pretty" to me. Japanese women had very coarse black hair, and this girl had probably never been that close to an American woman's hair. She probably could not resist touching it to see how it felt.

Barney's orders read that we were to return to San Francisco via MSTS (Military Sea Transportation System), a Navy ship designed to transport troops. I think we left from Yokosuka, but I have no memory of how we traveled there. To our discontent, we were to spend Christmas aboard, and there were no kennels for dogs. For Spook, we had to have a crate made which was chained on the deck. We had to walk him there and clean after him daily when the weather permitted. Unfortunately the ship endured some severe storms, and poor Spook developed pneumonia which delayed us in San Francisco where he was housed at a veterinarian clinic to heal.

Sailing across the Pacific in an MSTS was no joy, and I learned another Navy term called "titivate ship" (spruce it up). Our cabins basically were upper and lower bunks. Every day we were turned out of them while the quarters were being cleaned by stewards. Sometimes there was no room to sit in chairs, so we all sat crowded together on the decks. It certainly didn't make any difference then whether or not one was enlisted or an officer, although of course we dined in the Officers' Mess.

Barney, being a straight-laced naval officer, said there was no way he could take a bottle of alcohol aboard an American Navy vessel. That meant no Christmas cheer in that department. One of the Navy wives decided that rule didn't apply to her since she was traveling alone, so she generously shared a rum and Coca Cola with us on Christmas Day from a bottle she had smuggled aboard. I had made a little Christmas tree out of coat hangers for our cabin, and we exchanged gifts between each other. We thought that two weeks aboard a ship without much to do might be a good time to start a family. Unfortunately that didn't work, and we spent our entire married life with dogs for children. After that trip, I vowed I would never cross the Pacific again by ship and never again board an MSTS. At age 81 I have done neither.

While Spook was recuperating in San Francisco, we picked up our Buick and drove down to spend New Year's Eve at the Officers' Club with close friends stationed at Port Hueneme. It was a celebration to end our Far East adventure.

There are some personal experiences I have omitted from this tale that are best left unsaid. Captain Barney Martin and I enjoyed over forty-two years of marriage until he passed away on 25 April 1996. We designed and built our dream house in Rancho Santa Fe, CA, after Barney retired from the Navy in 1976 and furnished parts of it with most of the treasures we collected during our years in the Philippines and Japan.

We have cherished the memories and friendships from those exciting times, and were honored to have held one of the VW-3 Det A/VQ-1 reunions in our home. In retrospect, the fondest memories I have of our VW-3 Det A/VQ-1 days were of the people we met and shared lives. Our present reunions prove what an outstanding group we were, and our lasting friendships a valuable asset to us all. Those of us now left, and yes, the group is dwindling, still try to keep in touch. It was a closely knit Squadron, and due to our husbands' clandestine activities, we wives were always there for one another. My first experience as a "Navy Wife" was undoubtedly my best.

Jinny Martin

Autobiography

February 26, 2011

I spent my early years in Ohio and graduated from Miami of Ohio with a BFA. I worked as a graphic artist in Medina, OH. In 1950 I became an American Airlines stewardess, based in Tulsa, OK.

In 1954, I married Barney Martin, a naval aviator who was sent to naval squadron VW3-Det A in the Philippines. In June 1955 the squadron became naval squadron VQ-1. In Oct. of that year VQ-1 moved to Japan. We spent the first year of our marriage living in the Philippines and Japan.

Late in 1955, Barney became Naval Specialist Officer (Intelligence). For the next 21 years this new career designation led him to duty at the Pentagon, Europe. the Mideast, and North Africa. I resided with him at these duty stations except when he was the intelligence Officer aboard ship. During that time I traveled extensively, following the ship and maintaining residences where it put into port.

When Barney retired from naval service in 1976, we moved to Rancho Santa Fe, CA. Unfortunately, Barney died of leukemia in 1996. I still live there.

I enjoy animals and have been especially interested in the San Diego Zoological Society. I served for many years on The Wild Animal Park Advisory Board that promotes all aspect of the society in particular, and wildlife in general.

My passion widens to include animals in general and especially dogs, who've always been loving companions. I have accumulated a large library devoted to genealogy and family history. My interests also include horticulture, gardening and reading.

Lavon North

THE CALLING OF A NAVY WIFE:
Memories of LaVon North's First Navy Tour of Duty

LaVon, Dean and Gregg

LaVon North died on 11 January 2006. Her memories, as written here, were penned by her husband, Dean North, based on the knowledge and feelings she shared with him. Any errors of fact, time, or place are born of his misconceptions or his faulty memory, for which he humbly asks your forgiveness.

Our First Set of Orders

I remember how proud I was — how proud he was — as I pinned Navy Wings of Gold on Lieutenant, Junior Grade, Dean B. North, my husband. It was 30 November 1954 at the Naval Air Station (NAS) Hutchinson, Kansas. We were both glad that the long eighteen months of flight training were over and were excited at the orders Dean received for his first tour of flying duty. He, along with four friends, had put in their "wish lists" for patrol squadron duty in Hawaii. When the orders were posted, Dean saw all of his friends' names on orders to VP-22 in Hawaii, just what he had hoped for, but was dismayed when he saw his name was not included. Searching through the many sets of orders posted before his, he finally found a set with only his name on them thumbtacked on an obscure lower corner of the board. The orders instructed him to "Proceed to San Francisco, California, and report to the Commandant of the TWELFTH Naval District for transportation to the port in which Airborne Early Warning Squadron THREE (DETACHMENT ABLE) (VW-3 Det A) may be and upon arrival report to the Commanding officer, VW-3 Det A, for duty in a flying status involving operational

or training flights." He came home disappointed that we were not going to Hawaii with his friends and puzzled as to where we were being sent, and what kind of flying was involved. But, the next day he learned that a Commander Dunn, the Personnel Officer at NAS Hutchinson, had served in that Squadron. Dean talked with him and learned that the Squadron was very small; that although the parent Squadron was stationed in Guam, Detachment Able was an independent group based at NAS Sangley Point, a small peninsula of land jutting from the southern shore of Manila Bay in the Philippine Islands; that dependents were allowed to go there; and that he would learn more about the Squadron's mission later, for it was Top Secret. Dean came home all excited about such an intriguing and mysterious assignment. We didn't know what the future held, but we were both eager to launch into this new adventure.

But I was more than a little apprehensive as well. I grew up on Warnock Avenue in Salt Lake City as an only child named Margaret LaVon Ward. My mother's name was also Margaret, so I went by my middle name. I lived in the same house (the house my mother had grown up in) from the time I was two years old until I was twenty-two, when I married Dean. My Scottish mother had doted on me but also lived her life through me, so had given me little chance to be independent. When Dean joined the Navy, I was thrilled to leave Warnock Avenue, but wasn't well prepared for the life of a Navy wife whose husband would be away so much of the time. And we had our two-year old son Gregg to care for, plus a black cocker spaniel named Pokey. Just how this would all work out concerned me a lot.

We were fortunate to have leave over Christmas. After spending the holidays with our parents, Dean reported for a four-week training course at the Fleet Airborne Electronics Training Unit at NAS North Island, San Diego, California. Gregg and I went with him, staying for a month at the Ebb Tide Motel below Point Loma. While there, Dean did the paperwork to permit Gregg and I to go to the Philippines, but found it would be about a six-month wait for passage. So Dean took me back to my parents' home in Salt Lake City. I was disappointed about the impending separation, but welcomed the certainty of where Gregg and I would be until Dean learned more about what we were in for overseas. Before Dean left, we all got our medical shots to protect us from various diseases of the Far East. Upon wincing from his first shot, Gregg exclaimed to the nurse, "Mary, you give hot shots!" A few days later when we returned for more shots, the prospect was too much for Gregg. He bolted from the doctor's office and ran down the long hallway. Dean pursued and caught him just before he reached the doors to the elevator.

It was now early February. We said goodbye — and Dean was gone. The next I heard from him was when flowers arrived on Valentine's Day. He sent them from San Francisco before going to Travis Air Force Base for a flight across the Pacific.

Voyage Across the Pacific

The next six months passed slowly. My dad filled in as a dad for Gregg, and they formed a close bond that was good for both of them. Letters came from the Philippines. Dean told of the oppressive heat and humidity and of the intense rain. He told of the Quonset hut he lived in (the Bachelor Officers' Quarters) where the closets had light bulbs always burning near their floors to help prevent the spread of mildew on shoes and clothing. He told of the squalor of Cavite, the town just outside the gate of NAS Sangley Point, where garbage was merely thrown into the street. He wrote of nipa huts, of "taxi" riding in converted World War II Jeeps called jeepneys, of bare-bottomed kids running in the narrow streets, of fierce-looking water buffalo pulling wagons on dusty roads, of a half-formed chicken embryo still-in-its egg delicacy called balut, of the difficulty of understanding the Tagalog language, of getting blood on his shirt from sitting too close at a cockfight area. It all sounded so foreign, so intimidating. He wrote of renting a housing unit for us at a place called PhilAmLife Court near the checkpoint of Cavite, which was a station to control access to the town. I wondered how clean the place would be and why the people needed to control access to their town.

But the time did pass. We received our passport (Gregg and I were both on one of them), and then the paperwork finally arrived from the Twelfth Naval District for our passage and for our car, a green 1954 four-door Chevrolet which Dean had decided to ship. We were assigned passage on the USNS Barrett, departing on 25 July from San Francisco.

My parents agreed to keep Pokey, then drove Gregg and me from Salt Lake City to the Naval Supply Center at Oakland, California, to drop off the car for shipment. That accomplished, we proceeded across the bay to the docks at the Presidio in San Francisco.

The USNS *Barrett* arrives in Manila with dependents manning the rail.

It was there I met Nancy — Nancy Grau — a newly married wife also going to join her husband Gary, who was in the same Squadron as Dean. She had contacted me, and we had arranged to meet in San Francisco. I introduced her to my parents and Gregg; then we went to a restaurant for lunch. When it was time to say goodbye before boarding the ship berthed at dockside, Gregg became quite upset. When he realized he was leaving his grandpa and grandma, he clutched at his grandpa and pleaded, "Sompa, I don't want to go to the Philistines!" This brought tears from my mother and the "goodbyes" quickly became very emotional. Their little girl was going away from them for a long time — and with their precious grandson. I was very nervous and feeling rather numb. The experience had become more a dream than real.

I relaxed some after the ship slipped away into the bay. Now all I had to worry about was Gregg and myself. The ship slowly turned toward the Golden Gate Bridge. I had always wanted to sail under it and found a good vantage point. But as we neared the bridge, Gregg announced that he had to go — now! I took his hand and raced to the nearest bathroom. So as we sailed under the Golden Gate Bridge, I was inside helping Gregg answer a call of nature.

Gregg and I shared a stateroom with a woman and her daughter. The woman was not friendly and had an accusative tone in her voice. I felt very alone and uncomfortable: the voyage held little promise of being enjoyable. I was so glad to have the company of Nancy and wished we had been assigned to the same stateroom. That night as I tried to fall asleep, I thought of the events of life that had led me to this uncertain journey.

I was eighteen when a new family moved in next door to our house. They had a son my age and a Springer Spaniel dog named Jerry. Jerry came over and introduced himself in exchange for a handout, but for over a year I never met the son. As I learned later, he attended the University of Utah during the day and worked at United Air Lines during the night. As I sat in church one Sunday in August 1947, a young man approached me and introduced himself. He said his name was Dean North. He asked me where I lived and was surprised to find I lived next door to him. He then asked me out on a date. I learned he was from Wisconsin, that he had an older sister who was on a mission for our church, and that they had moved to Utah after joining my church, the Church of Jesus Christ of Latter-day Saints. We just seemed to be right for each other. We went to movies and canyon parties. We went swimming in the Great Salt Lake and danced to the big bands of the forties that came traveling through Salt Lake City on their way east or west. After almost three years of going together, we were married on 1 May 1950.

In 1951, with the Korean Conflict raging, a letter from the Selective Service arrived in our mail. It directed Dean to report to Fort Douglas (located on Salt Lake City's east bench) for his draft physical. At Fort Douglas, they told him if his wife was pregnant, he would be deferred. I was pregnant, but didn't know it at the time. To avoid being drafted into the Marine Corps, Dean joined the Navy. I moved back with my parents. And so began our military life — apart.

After boot camp, Dean went to Officer's Candidate School in Newport, Rhode Island. I knew he would not be with me when our first child would be born, due in July. As the time approached, I became more nervous and worried. My doctor asked if he should get my husband home. Tears rolled down my cheeks as I summoned the courage to tell him, "No!" My mother took Dean's place when Gregg was born and sent him a telegram telling him he now had a son.

Dean received his commission as an Ensign in October 1952. Leaving Gregg with my parents, I flew out for the commissioning ceremony, after which we took a ten-day second honeymoon trip (the first one being a week in Las Vegas after our wedding). We saw the stage play "The King and I" in New York. We rode in a horse-drawn Hansom Cab around Central Park, we drove down the coast to Georgia where we rode in a canoe through the Okefenokee Swamps, and then to Florida where we rode the glass-bottomed boats at Silver Springs, and finally, at Sea World, I was selected from the crowd to feed the porpoises. After a year of separation, we were together again, even though it was for a short time.

I flew home to my parents' house while Dean attended Aviation Ground Officers School in Jacksonville, Florida. He drove home for Christmas, arriving at my parents' house on Christmas Eve. I met him at the door and led him by the hand to our bedroom where, in a crib, a little guy lay sleeping on his tummy, his fanny raised in the air. Arms around each other, I presented him with our son, Gregg — the three of us together for the first time.

After Christmas, we all went to NAS Memphis where Dean attended Aviation Electronics Officers School. There he met several aviators back from the Korean conflict. Dean had learned to fly as a civilian. (He and nine others owned an airplane when he was eighteen. Later Dean sold his share to buy me an engagement ring.) One aviator took him flying and encouraged him to apply for Navy flight training. Dean asked me what I thought. Perhaps it was his selling the airplane to buy my engagement ring — perhaps it was his enthusiasm for flying — I don't know, but I told him he should do it. He applied for flight training and was accepted, a decision that then led to months of training in Florida and Kansas and had now put me on a ship headed for a faraway place that sounded dreadful.

But Gregg and I soon fell into the routine of shipboard life. Fortunately the weather was good all the way, so I wasn't seasick. I spent much of each day with Nancy. We had lots in common and became close friends, a friendship that would continue for a lifetime. We sat and talked, we saw some not-too-good movies, and we sunbathed. Each day the passing of time was marked by the meal calls dividing morning from afternoon from evening. Nancy had no knowledge of being a Navy wife so was interested in my experiences over the past four years. We also met the new Commanding Officer of NAS Sangley Point and his wife. She was a friendly person with no air of superiority.

Gregg's third birthday was July 27th, just two days after we set sail. The ship hosted a birthday party for all who had birthdays during the crossing. I told Gregg the party was for him alone. After all, how many kids get to have a birthday at sea?

When we crossed the International Date Line, we all received our "Shellback" certificates and lost a day of that week. After thirteen days of sailing, we arrived at Agana, Guam, for an overnight stop. Wanting to get off the ship for a change, Nancy, Gregg, and I took the opportunity to go on a sightseeing tour. Most of what we saw was Anderson Air Force Base. We did have some time to go shopping in some stores nearby. There we bought matching Aloha shirts for our guys and us. Then we returned to the ship. The next morning we began the last leg of our journey. It occurred to me that I now had more time at sea than did my Navy Officer husband!

Three days later we watched as a mass of land ahead of the ship grew ever larger — it was our first sight of the Philippine Islands. We slid by many islands and land masses as we made our way around to the western side of the main island of Luzon. However, it was the next day before the broad entrance to Manila Bay came into view. As we passed the island of Corregidor and the Bataan Peninsula, both made famous in World War II for the battles and forced marches of prisoners that took place there, we could see the city of Manila taking shape ahead. Then the ship slowed as a small boat approached. It was the pilot boat bringing out the harbor pilot to take the ship safely to its berth. Later, after we had safely traversed Manila Bay and drew near to Manila, the same boat came out to pick up the pilot. As we looked down at it, Nancy pointed and excitedly exclaimed, "That's Gary down there on that boat!" I looked down and saw Dean beside Gary, waving at me from below. I pointed him out to Gregg, hoping he hadn't forgotten his father's face. If he had, I couldn't blame him, for Dean had been gone eighteen months of the first three years of Gregg's life.

As the pilot boat pulled away, the ship proceeded to the pier at an agonizingly slow pace. Nearing the dock, Nancy, Gregg, and I were at the railing looking down at the hundreds of people gathered below. Then Nancy saw Dean and pointed where to look. I saw him and then found Gary, both waving at us. Gregg saw them and waved back. Waves of relief and emotion swept over me. I had been alone so much: I needed someone to lean on, someone to take the lead, someone to reassure me that all was well. Soon his arms were around me. We looked in each other's eyes, and we kissed. Then Dean swept up Gregg and gave him a hug. Gregg wasn't really sure who that big man was and quickly reached out to me. But we were a family again. The long wait was over!

It took over an hour to get our baggage unloaded and claimed. Then we went to the U.S. Embassy boat landing and boarded a boat for Sangley Point. As we crossed Manila Bay, Dean and Gary pointed out many things to see, but I was so caught up in just completing the ocean voyage I saw and heard little of the strange sights, sounds, and smells surrounding me in this country so foreign to my senses. At Sangley Point, we went to the Officers' Club for dinner. The modest club was housed in a Quonset hut, but the food was good. Then Gary took Nancy to their house in Cavite, and Dean took Gregg and me to our new home in PhilAmLife Court. It was nice to see our furniture again. Dean had had some of our stuff shipped out before I arrived. As my anxiety diminished, I became very tired. My long journey was over.

Life in Cavite

Our home at PhilAmLife Court was certainly tropical. My skin never dried but remained slippery from the humidity. And a permanent odor permeated the air, like wet dog hair mixed with the smell of food fried in rancid oil. There was no escaping either the humidity or the smell because the window openings of our housing unit contained no glass, only sort of shutters open to the outside. Our unit was one of four in one building. We had two floors: a small kitchen and living space below and two bedrooms and a bath above, all of which begged for a good scrubbing and some paint. Outside were only sidewalks, dirt, and a street. The court contained six or eight buildings like ours and was walled in on all sides except for one gate, at which was posted an armed guard. The problem was thieves — both two-legged and four-legged. The high walls surrounding the court were festooned across the top with glass shards embedded in cement. On the other side of the wall behind our unit were broad pools of sea water. They were part of a salt farm, which reclaimed sea salt for commercial use. And we soon became acquainted with Finnegan, a battle-scarred junkyard dog that "owned" our court due to his driving away or killing all other dogs and cats that dared venture into his kingdom. We trusted Finnegan more than we did the armed guard, for everybody in the court fed Finnegan. In the middle of one night, we were awakened by a howling, screeching fight outside our window. When we arose in the morning, we found that Finnegan had killed a cat in our small front yard.

PhilAmLife Court was on the opposite side of town from NAS Sangley Point. Our car had not yet arrived. Seeing the twisting, narrow streets congested with animals, chickens, kids, and garbage, I wondered why Dean had shipped the car in the first place. We learned to board the Jeepneys for transportation to the Base. The highly modified World War II jeeps were decorated with brilliant, sometimes garish, paint jobs, often with a religious tone to them. Each one was different, expressing the personality of its owner. We entered the open-air seating area through the rear of the vehicle, sitting on bench seats along each side. The fare was ten centavos regardless of where you were going, but it wasn't nonstop. If someone got aboard later and was going to a closer, but out-of-the way place, the Jeepney turned and delivered that person first. Breakdowns or just running out of gas were not unusual. For more dependable transportation, Dean bought a Lambretta motor scooter at the Navy Exchange. The seat had room for me to sit behind Dean while Gregg could stand between Dean's legs — and away we would go. Finally our 1954 Chevrolet arrived, but it was of little use. We should never have brought it with us.

Gregg loved to watch the geckos run up the walls and across the ceiling of our unit. They kept the bug infestation in check, so were welcome residents. Gregg loved to play "the dead bug game." During the night the geckos would catch roaches and other bugs and take them under the kitchen sink to suck out their juices, leaving just a dry carcass behind. Each morning Gregg would run downstairs to open the doors below the kitchen sink and count the night's kill. Our Squadron friends loved to ask Gregg to play the dead bug game, upon which he would roll on his back and thrust his feet and arms into the air.

Gregg in his Barong Tagalog — they called him "Magsaysay Boy."

The months of apprehension and separation had taken their toll on me. I had lost a lot of weight and never really felt well. It grew worse in Cavite. The oppressive heat, the drenching humidity, the food, the water — all taken together were more than my system could handle. I became so sick and thin I couldn't even care for Gregg properly. Gregg was not doing well either. His fair skin burned and peeled in the blistering sun and the humidity kept him from sleeping since he was always wet with sweat.

We hired a maid, Ramee, to do the housework and help with Gregg. And Dean installed an air conditioning unit in Gregg's room (one he had had shipped out from the States for use in the BOQ). It knocked down the humidity, even though the room was open to the outside. Ramee washed our clothes in the bathtub and polished our wood floors by standing with half of a coconut under each foot and scooting around the room.

I finally went to the Base doctor for help with my illness. He had me get a chest X-ray. A day or so later, I was informed that I had a spot on my lung. I was frightened and dismayed — why did I ever come to this place? Then, a day after that, they called and said they noticed that everybody who had a chest X-ray that day had a spot in the same place on their lung. It was some dirt in the machine! I started feeling better — nothing like a good scare for a cure!

Soon I met the other wives, many of whom became lifelong friends. Some lived in Base housing, others lived in Cavite. Joan Angier was always "up" and cheerful. Dean said that her husband Don was one of the best pilots he had flown with and that when Don was around, times were never dull. Margaret Burdg, quiet and sincere, welcomed me with kindness and support. Dean told me that he and her husband Don had tried to go to India on an Embassy Flight. But after Don accomplished the mountain of required paperwork, after securing all the permissions, after obtaining visas, and after having leave papers in hand, the Skipper called them in two days before they were to depart and canceled their leave (Dean was not happy about that!) I met Dottie Flood, a Canadian, newly wedded to her husband Frank. And I met Pauly Varner who was most helpful during my time in the Philippines. Dean was on her husband's flight crew. They called him "Stretch." Dean had nothing but respect and praise for him.

To my surprise, I saw a familiar face among the Squadron wives — Helen Ebersbaker. She and her husband John had been "across the hall" neighbors in our apartment building at Millington, Tennessee, while Dean was attending the Aviation Electronics Officers School at NAS Memphis. John was a non-pilot officer. Dean told me each flight crew had one. And I met Elizabeth Rathbun, wife of Wells, another non-pilot officer. We called her Liz and I immediately took a liking to her. And there were Jinny and Barney Martin, Bea and Joe Laubach, Jean and Bob Hanson, and Alice Sumner, also married to a Don (are all the husbands named Don?) And I met Lee Chapola, also a wife of one of the non-pilot officers. I soon found out firsthand that her husband Pete was truly an outstanding dancer. I also met Cathy Hall, wife of Eugene Hall, Commanding Officer of VQ-1, and others — too many for me to grasp all at once.

Not surprisingly, the Squadron wives were socially active. I didn't care to play cards and didn't particularly enjoy cocktail parties, plus I had Gregg to take care of, so I declined some kind invitations from the other wives. I hope they didn't consider me standoffish. But there also were dances, outdoor barbecues, trips to Manila, and other get-togethers that were great fun.

The Base at Sangley Point was not large, mainly a runway and a seaplane ramp with some buildings placed where space was available. The foliage was abundant, bearing huge flowers of outrageous colors. The Base had a swimming pool where Gregg could splash and dunk with many other kids, although the pool would often close without notice to take care of some kid's "accidental discharge" in the water. The Navy Exchange was open for us except on Wednesdays when it was reserved for Filipinos. We irreverently called it "HUK Day," after the rebel Hukbalahap group that terrorized the countryside. It seemed to me the merchandise was geared more to Filipinos than anyone else. For example, they only carried wide shoes (I wore narrow). The Officers' Club was a Quonset hut near the runway. A short distance away was another Quonset hut housing the Bachelor Officers Quarters (BOQ), where Dean had lived for the past six months.

Cavite was unbelievable. Nipa huts, basically a loose collection of wood and nipa palm thatch up on stilts, and all other manner of basic dwellings perched beside stagnant standing water or along dirt streets lined with garbage, that people threw from their yards. Children seemed to fall into two categories: those not potty trained who wore shirts but no pants, and those who had graduated from that state who wore pants but no shirts. They played in the filth of the town. No wonder the local casket shop had more small caskets on display than full-sized ones. Shops were open to the air, selling food, clothing, trinkets, furnishings, and whatever. The marketplace held an unappetizing collection of food and was located on a dirt road under a canopy of dirty canvas covers held up on sticks to provide protection from rain. The collective odor was certainly memorable: a stench of rotting garbage, stale food, and pungent fish. Flies landing on the bloody mess being displayed seemed to concern neither the vendor nor the customer. I wondered how long ago any food handler here had washed his hands. And the staccato bursts of Tagalog were beyond comprehension, much less to be learned and spoken, so I bought very little in Cavite. The peso was the standard of exchange, but to reduce black market exchange, many shops would take Military Payment Certificate (MPC) that we were paid with instead of dollars. It was risky for the shop owners because the military would periodically close the gates to the Base without notice and exchange the current MPC for a new issue that made the old MPC worthless.

We could buy huge avocados for twenty centavos that would make a lunch for both Dean and me. My favorite drink was calamensi juice, made from a sort of miniature orange or tangerine citrus fruit. And the Officers' Club served delicious Langouste, a clawless lobster found in the waters of the South China Sea. My favorite Filipino food was Lumpia, a small version of a Spring roll made with finely chopped chicken or pork and vegetables rolled into a small pastry and fried. It is served with a sweet dipping sauce.

The Navy operated a boat that traveled between Sangley Point and the American Embassy at Manila. This was the accepted mode of transportation to Manila for military personnel and dependents. Road travel was too dangerous. As we rode across Manila Bay, it was still littered with the half-sunken wreckage of many ships from the battles of World War II, their rusting hulls and superstructures emerging from the water in grotesque shapes against the sky. At Manila we visited the former Japanese internment camp at Santo Tomas University. The stone cells and narrow passages that had held prisoners were now squalid shelters for families living in desperate poverty, their naked children playing in filth. We visited Malacanyan Palace, the home of Philippine President Ramon Magsaysay, a place of grandeur and wealth. The abrupt change from Santo Tomas was stunning, an instant trip from the bottom to the top of Filipino society. Manila seemed to be like that - bipolar. The stone mansions of the very rich towered over cardboard shelters and caves that were home to the very poor. Mercedes cars wound their way through groups of dirt-covered children playing in the street. Opulent tombs rose skyward beside torn earthen grave mounds open to the animals and elements. As we walked, a brown-skinned boy clad in only a ragged shirt ran up to Dean and shouted, "Hey, Joe, gimmee centavo," while his streetwise friends hovered nearby, ready to rush in if any success was in the offing. It was obvious they considered Americans an easy mark.

But there was a better side to a trip to Manila. The Embassy served great lunches. We browsed the shops and art stores along A. Mabini Street where we bought an oil painting of a fishing boat in Manila Bay at night. We learned of beautiful Filipino handicrafts in cloth, silver, and wood. At a Catholic church we saw an organ whose pipes were made of bamboo. While there we bought a silver filigree spoon so intricate it appeared to be lace. And everywhere people would come up and ask if they could touch Gregg's sun-bleached

Dean North in Formosa (now Taiwan)

blonde hair, an oddity in a land of only black hair.

So this was my world when Dean was away. I never worried about his flying. I knew he was a good pilot and I knew he would not take unnecessary risks. Perhaps if I'd known what their mission was, I wouldn't have been so trusting. Dean would leave early each day because, due to the scorching afternoon sun, the Squadron operated on tropical hours (0600 to 1200). Sometimes he would say goodbye without notice and tell me he'd be back, but he couldn't say when. But I thought it was interesting that when I packed his bag, he'd tell me how many pairs of underwear he'd need, so I usually had a good idea of how long he'd be gone. I didn't venture going to the Base when I was alone. I didn't like being left on my own in such a strange environment, but having been an only child and having Dean gone for over two of our five years of marriage, I was accustomed to being by myself, and I had Gregg for company. Perhaps I was learning to be independent and rely on myself, the calling of a Navy wife.

Dean would bring me presents from each trip away, mostly presents from Japan: delicate fans, slippers called zories, a black pearl necklace and earrings, a jacket for Gregg with embroidered Japanese designs on the back, beautiful vases and silk paintings. Each week I felt more at home, more at ease with my new life in this strange land.

One evening Dean called me to come outside of our quarters at PhilAmLife Court. He pointed at the sky where many circles of flickering flame were drifting along high above our heads. We wondered what they were. Recent rumors of flying saucers from all over the world came to mind. Claims of seeing circular lighted objects in the skies, some in daylight and others during the night, sometimes stationary, then moving away at unbelievable speed, had been reported from many countries. We felt that was science fantasy, yet here were what seemed to be actual saucers flying over Cavite. Dean called Base Operations at NAS Sangley Point to report what he was seeing. They called Clark Air Force Base to scan the area with radar. They reported no contacts above Cavite. Just then one of the circles of flame billowed up into a large fire and began descending ever more rapidly, crashing into the ground just outside our housing compound. Dean, along with others, rushed over and found a large, shallow, metal pan containing flaming oil tied beneath a paper hot-air balloon. It had crashed because the paper balloon had caught on fire. A nearby Filipino told Dean that this contraption had been launched as part of a religious festival on the shore of Manila Bay. Dean reported this to Base Operations who later called back to thank him because the flaming balloons had begun plummeting down into the Naval Air Station munitions yard.

We often went to activities in the company of Nancy and Gary Grau. Gary was fun, always smiling, and he always had some comment ready for any turn of events. One weekend we four and Gregg traveled in our car south from Manila Bay to a resort high above Lake Taal. The view was worth the ride. Lake Taal extended for miles, surrounded by ragged hills rising from the water's surface under hazy summer skies. Jutting up from the lake was a volcano cone in which was a deep crater nearly filled with water. In the center of this crater lake rose a vertical bread-loaf-shaped island of rock. And so the saying went that the rock was an island in a lake in an island in a lake in an island in a lake, the last lake being the Pacific Ocean. We took pictures, including a favorite of mine — Gregg, Dean, and I standing on a hill high above Lake Taal.

Glen Carlstrom, a young Squadron pilot, came to the Philippines a bachelor, but would not leave as one. He had met a young lady named Marilyn whose stepfather was an official at the U.S. Embassy in Manila. Glen proposed - Marilyn accepted. The wedding was a big event for everyone. After all, there was only one other bachelor pilot in the Squadron. And how much more romantic can you get than being married beside the South China Sea? The men would be in their dress white uniforms. We wives searched our wardrobes for what to wear. Dean and I bought Gregg a Barong Tagalog to wear. The Barong Tagalog was the accepted formal wear for men in the hot, humid Philippines and one of the best ideas I've seen. It had no tie, no coat, and wasn't even tucked in the pants. It was very delicate, being made of pina-fiber see-through fabric and embroidered with intricate patterns across an area on the front of the shirt. Gregg's Barong Tagalog was a hit at the wedding.

My thoughts also go back to two days before the wedding, when Marilyn happened to see Dean and me as we waited at the Embassy dock for the boat to Sangley Point. Marilyn came up to talk to us. She was nervous and worried; she had the pre-wedding jitters. She wondered if she was doing the right thing; she wasn't sure she was up to all the celebration and attention. She asked if I had felt the same; way before I married Dean. I reassured her that what she felt was normal for a bride. I told her of my apprehension, my fears, and my short temper during the final forty-eight hours before the ceremony. And I told her how all that melted away once the ceremony began, how thrilled I was at the reception, and how I have never regretted my marriage. Her countenance brightened with relief and confidence. (I never shared this story with Glen.)

There was also a wedding we missed. Back in my hometown of Salt Lake City, Dean's sister married Kenneth Judd in August 1955. Between Dean and me, she was our only sibling. There would be no more weddings in that generation of our family. Navy life exacts a price, a sacrifice of missing once-in-a-lifetime family events. Dean wasn't there when I gave birth to his son, he wasn't there for most of Gregg's first three years, we had just been apart for another six months, and now we weren't there for his sister's wedding.

A Second Set of Orders

On June 1st, 1955, before I arrived on the scene, the Squadron had been commissioned as Electronic Countermeasures Squadron One (VQ-1). Now we were told that the newly formed Squadron's home Base would be moved to the Naval Air Facility (NAF) Iwakuni, Japan. The move would take place in the fall, so my stay in the Philippines would last only two months. There were places I had hoped to go, such as Baguio and Hong Kong, but would now have no time. Yet Japan sounded like an exciting place to live — and we'd be away from the heat and humidity as well.

To prepare for the move, we decided to sell much of our household belongings, including our 1954 Chevrolet for which there was a ready market. We sold it for many hundreds of dollars more than we had paid for it, all paid to us in Filipino currency — pesos — at an exchange rate of two pesos to the dollar. The other Squadron members sold similar items. Soon we collectively had many thousands of pesos. But pesos would be worthless in Japan. We needed to exchange them quickly for MPC. Just how to do that was the problem.

To start this process, the Squadron set up a money exchange booth by the main gate of the Base where the sailors could buy pesos when going on liberty. That worked to an extent, but many thousands of pesos remained. NAS Sangley Point was small compared with the huge Navy base at Subic Bay on the northwest coast of the Bataan Peninsula. Someone suggested that we would have much better success there in exchanging our pesos. So Don Burdg put some $10,000 worth of pesos in an SNJ (a single-engine airplane used by the Navy for training) and flew to Subic Bay on payday. We all crossed our fingers that the plane would make it

A Torii on the shores of the Inland Sea

there and back safely and that Don would return with our small fortunes intact and converted to MPC. It did, and he did!

Dean told me of the plans that had been made for the move to Japan. An advance party had been sent to set up the Squadron spaces and arrange for dependent housing. Squadron equipment and household effects would be sent by ship. Then during late September, the flight crews would fly up in the Squadron aircraft, and the dependents would be flown up by a Navy Fleet Logistic Air Wing (FLOGWING) aircraft. At a Squadron party for a departing officer and his wife, Cathy Hall came up to Dean and told him he would be leaving for Japan with Stretch Varner's plane on 30 September, and that LaVon would leave that same day with other dependents. Dean was not pleased that the first word he heard of this came from the Skipper's wife. How did she become a part of the chain of command, anyway?

But we learned her information was correct. And we learned that many of the Squadron personnel, including us, would be quartered in Army housing at a place called Niji Mura, located some twenty-five miles from NAF Iwakuni across the Inland Sea of Japan. The men would have to commute to Iwakuni each day by boat.

On 30 September, Dean, Gregg, and I went to the Base for the "big day." Our belongings, except for Dean's B-4 bag and my two suitcases, had previously been crated up for shipment. Soon Dean's flight took off. We were waiting for further information on our airplane when it was announced that our flight had been canceled! What? How could that be? We were told that a major typhoon had swept over southern Japan and damaged the housing at Niji Mura. Beyond that, the Navy had little information for us. How bad was the housing damage? How long before we could expect to leave? When would we be given more information? No one knew. We had been deserted. We were left homeless in the Philippines.

The wives "circled the wagons," as it were, and came up with a plan. As part of that plan, Pauly Varner took Nancy, Gregg, and me to her quarters on Base, which were still available to her. Once there, I tried to open our suitcases and found that Dean had locked them — and had taken the keys with him! The day was not going well at all.

We thought we would probably leave in a couple of days, but that was not to happen. Each morning we'd arise, hoping our flight would be called, but days passed with no word on our departure.

I learned that Dean had spent two days in Okinawa, but was now waiting at NAF Iwakuni. Pauly and Nancy kept my spirits up, but my patience with the Navy was wearing thin.

The day finally arrived that we were notified that a FLOGWING DC-4 aircraft had arrived to take all of us to Japan — wives, kids, and some of the Squadron personnel. Nevertheless, after being told to board the aircraft, they changed their minds and we were told to turn around and return to the Snack Bar. It seems they had loaded the baggage without weighing it — a necessary item in determining if we would be too heavy for takeoff. We all went to the "Geedunk" (a term for the Navy Exchange snack bar) and sat around for over an hour wondering if we would ever see our husbands again and speculated as to what more bad luck could befall us before leaving the Philippines. Finally, we were told to board once again. This time all went well and soon we were airborne. Sangley Point grew smaller to our view, and then disappeared from sight. I was so relieved that we were on our way at last.

We sat in canvas-covered bucket seats along the sides of the airplane — definitely not first class. A constant vibration ran through the seats and airframe, and the engine noise made conversation difficult at best. As we bounced around in the tropical air currents, I held on to the thought that if Gregg and I could just survive this one day, all would be well. I should have suspected that the test of our endurance was just beginning.

The Longest Day

Gregg quickly fell asleep to the droning of engines and the gentle rocking motion of our aircraft in flight. Facing inward, there was little for me to watch, just mothers and kids making the best of the situation. Drifting away in thought, I reflected on how I always seemed to be left behind, then would have to scramble to catch up with Dean. It was that way when he left for boot-camp. I stayed behind with my parents, then flew to San Diego to see him for just one weekend. After boot-camp he was sent to the East Coast and on to Officer Candidate School in Newport, Rhode Island, leaving me behind while I gave birth to our son. I flew out to see him at graduation time. Then he went ahead to Electronic Officers School at NAS Memphis, leaving me again with my parents until he found a place for us to live. It was a two-month wait before Gregg and I joined him. And it was the same way when he started Ground School in the Flight Training Command at Pensacola: Gregg and I finally caught up with him three months later. I thought the six-month wait to join him in the Philippines would result in more permanency, but now, after only seven weeks, I was once again scrambling to catch up with my husband.

Several hours later, my reverie ended as I heard the engine noise diminish and felt the plane begin to settle. We were on approach to NAS Kadena in Okinawa for a refueling stop. Gregg woke up crying. His ears were hurting from the increasing air pressure as we descended. I tried to tell him how to clear his ears but without success. He wasn't alone in that malady. The wailing of children continued throughout the descent and landing.

It felt good to walk around and restore circulation to our limbs. Somehow my mind had prepared me for just one flight, but we were now only halfway to Japan. I was tiring and Gregg was, too. I dreaded the thought of going through it all over again. But soon we were aboard once more, bouncing down the runway and into the air. Box lunches were passed out, but I wasn't in the mood for food. Gregg nibbled a bit, then fell asleep. My legs gradually went numb from the uncomfortable bucket seat, but somehow the hours passed.

Turbulent air brought me back to the present. The bouncing increased. Gregg was irritable when he awoke; he had had enough of this airplane. Then came more bad news. Word was passed that, although we were nearing Iwakuni, our landing would be delayed for some reason. Is this plane jinxed? We began circling endlessly through the overcast, waiting — and wondering — and wishing this trip was over. The mood of all aboard was sullen. I could hear some of the youngsters complaining and asking, "When do we get there, Mom?" and "I want to go home!"

After forty-five minutes, we finally started our approach to Iwakuni. Muffled sighs of relief echoed through the cabin. But nothing could match the expressions of gratitude (made in various forms of speech) that burst forth when we felt and heard the wheels scrunch down on landing. As the airplane slowed and turned off the runway, relief and a burst of new energy swept over me. We were finally there!

We deplaned into chilly air under a gray overcast, now darkening as the day's light faded away. It was now nearly six o'clock in the evening. Once again the cycle of another reunion with Dean brought hugs and kisses. I was ready to get on a bus, a car, anything, and go to our new quarters. My ears hadn't cleared from the flight, and all I wanted to do was rest. But that was not to be. Dean told me we would be taken across the Inland Sea of Japan by boat to our quarters at Niji Mura after we first had something to eat. Then he added that the boat ride would take two-and-one-half hours. I was ready to quit. Whoever planned this trip had no understanding of what it's like to travel with kids, but there was no other choice. I can't remember where we ate, but night had fallen when we were taken to the boat dock.

I couldn't believe my eyes! There was no "boat" for us to board; there was only a converted landing-craft with its nose on the shore and its flat ramp laid down for us to enter, like we were soldiers loading to assault a beach somewhere. Inside seats had been installed in the hold for us to sit on. With everyone in place, the diesel engines roared into life and, with the smell of diesel oil in our nostrils, we backed away from the shore, swung around, and headed into the pitch black void of the Inland Sea.

The flat-bottomed landing craft rolled and pitched with every wave. I was so tired my mind couldn't think any more. Fortunately, Gregg curled up and went to sleep. Time passed more slowly than I thought possible. There was nothing to see, nothing to do, nothing but try to keep upright as we rolled and rocked through the water. And it was cold sitting in the hold. The metal seat was drawing heat away faster than my body could replace it. Agonizing was too soft a term for this trip. But eventually we heard the engines reduce power and then the scrunch of metal on concrete as we beached on what we found out was an abandoned seaplane ramp. As we then learned, our Niji Mura housing had been built by the Australians on a Japanese airfield once used by Kamikaze pilots in World War II.

Our quarters were nearly a half-mile walk from the boat landing. It was after ten o'clock when we received our keys and arrived at our assigned house, one of four two-story units in one building. For military quarters, our unit seemed clean and sufficient and was far more to my liking than PhilAmLife Court had been. But our quarters were cold — there was no heat — and we were already chilled to the bone from that long, cold, boat ride. We were so weary and tired. Try as we might, we couldn't find how to turn on the heat. Then came another surprise. The Army (who ran Niji Mura) hadn't expected us until the next day. There was no heat in any of our quarters, and we were told there wouldn't be any until the next day.

We bundled up Gregg under all the blankets we could find. Dean turned all the lights on, hoping the heat from the bulbs would help. It didn't. We kept our coats on and piled on anything else we could find. We were all so exhausted we did get some rest, but the clock's hands moved ever so slowly as I longed for the dawn to break and bring some warmth from the sun. I found it distressingly fitting that the longest day in my life was now followed by the longest night.

What a Difference a Day Makes

As the sun rose, we gazed out our windows to see what Niji Mura looked like. It was beautiful. The housing units were painted cream with green trim. They were set back neatly along a hardtop road and each unit was surrounded with a green lawn. Small pine trees dotted the grounds, and each front walkway was lined with flowering plants. The tree-shrouded nearby hills appeared like a picture on an oriental postcard.

Venturing outside, we found the air was fresh and clear, although it brought with it a distinct odor different from the tropical musty smell of Cavite. The smell reminded me somewhat of the pulp paper processing plant near Pensacola when Dean was in Basic Flight Training. We learned that our unit was between Barney and Jinny Martin and Bob and Jean Hansen. Gary and Nancy Grau were assigned a unit in the building next to ours. We also located a small Post Exchange, a movie theater, and an Officers' Club. Being located well outside the city of Kure, Niji Mura had a quietness and peacefulness about it. And mid-morning, the heat came on. I immediately felt good about Niji Mura and our future there.

Our unit was basic, but comfortable: a living room and kitchen with an eating area downstairs and two bedrooms and bath upstairs. We had shipped our furniture from Cavite, so soon our home was furnished with familiar things. We hired a maid, Asiko Imade, whom Gregg called "Coco." She was taken with Gregg, and he was happy with her attention, which allowed me to come and go without worry. Our lawn and flowers were cared for by a gentle older man named Amano who came by one morning each week to surround us with the beauty of nature and blossoms. And with the Squadron wives now all living near one another, I had company wherever I went.

Life at Niji Mura

We quickly fell into a routine of living. Dean's work days were long. He left early each day to walk to the landing where the boat to Iwakuni left at 0530, arriv-

Bob Galloway & Glen Carlstrom on morning boat to Iwakuni from Nijimu.

ing there at 0800. At day's end, the boat left Iwakuni at 1700 (5:00 PM) to arrive at Niji Mura at 1930 (7:30 PM). Many times Gregg and I would walk to the boat landing to meet Dean as he arrived. We'd walk home where we'd have supper, sometimes very hastily, in order to catch the evening movie before retiring for sleep and another day. And when Dean was away flying missions for a few days, I was freer than I have ever been.

During the day, the wives would often go into town or just out for a ride. One of the wives, I believe it was Jinny Martin, had a Volkswagen "Bug" to get us around. What fun we had driving around Japan, mainly into Kure (the nearest city), where we enjoyed shopping, sightseeing, and eating. Some business names, in their attempt to "go American," were quite humorous. One such restaurant named "The Gastronome" proclaimed its main dish as a "cow filet."

The language was distinct. Soon I learned enough words to find my way, do basic shopping (including "make it more cheap"), and order food. I learned the greetings first: Ohayo gozaimasu (Good morning), Konnichi wa (Good afternoon), Konban wa (Good evening), Oyasumi nasai (Good night), and Sayonara (Goodbye). Then I mastered Hai (Yes), Nei (No), Gomen nasai (Excuse me), Domo arigato gozaimasu (Thank you very much), and how to count to ten. I never was ready for the fast lane of Japanese communication, but with my small vocabulary and the fact that the Japanese knew quite a bit of English, our verbal exchange was tolerable. Of course, I hadn't the slightest clue when it came to reading Japanese signs and labels. On the other hand, Gregg became almost conversant in simple Japanese through Coco. She watched him while I went out with the other wives, and she took him into town to shop. One day she took him to her home, where she introduced him to her beloved father and to her daughter. This was a special event for Coco and an honor for Gregg.

While shopping one day, I believe I introduced the layaway plan to Japan. I found a hibachi I liked very much and in bargaining as to the price (using one of the first Japanese phrases I learned, "Make it more cheap"), the Japanese store owner and I reached agreement. I found I did not bring enough yen to pay for the hibachi, so I asked him if I could pay half now, have him keep the hibachi overnight; then I would come the next day to pay the other half and take the hibachi home. He could not understand. It was fine with him that I pay only half now, but why would I want to leave the hibachi with him? He would trust me completely that I would come back to pay him the other half. I finally convinced him that he should hold the hibachi for me, but he was shaking his head and muttering when I left.

Some of the Niji Mura group organized an evening dinner party at a geisha house in Kure. Sitting low around a long show table in a beautiful room, we were served dinner by beautiful geisha girls dressed in stunning colorful kimonos. After dinner, they entertained us with songs and samisen music from a small tatami mat stage. Then they brought us into the fun with challenges to do better than they in certain dexterity tricks. They challenged Dean to roll up a long length of string onto a wooden spool faster than the geisha could. If he lost, he was to dance with the geisha on the stage. He accomplished the feat quite rapidly, in just a few seconds. Then the geisha put him to shame in a twinkling of an eye. Dean bravely went up to the stage where he danced to the tune of a Japanese song, "The Coal Miner's Dance." His lack of dancing skill was vividly apparent, but he was a good sport and we, including the geishas, all gave him a round of applause.

The Inland Sea of Japan bordered our housing area, its beauty surrounded by sculpted green hills topped by graceful silhouettes of trees. Gregg decided to make boats out of newspaper and then wanted to sail them on the Inland Sea. One evening we walked to its shore. The surface was calm with just the suggestion of small ripples. Gregg launched his boats. Of course, they immediately became waterlogged and partially sank, but sail them he did.

Cherry Blossoms near the Kentai Bridge in the spring

Landscaping and flower arranging in Japan is a high art form of beauty that captures one's imagination and speaks to the inner spirit. The smallest back yard can be a sanctuary of peace and stunning beauty, giving no impression of being confining. Several of us took a class in flower arranging to learn the secrets and skills of the art. I knew when the arrangement was right, but I wasn't very good at arriving at that point. However, some of the skills did rub off, and I have put them to use as best I could. The container or vase is part of beauty, and must be selected with care and an eye to the completed work. So I collected several dishes and vases that caught my eye with their aura of what is Japan.

Many other objects of Japanese art caught my eye and my pocketbook. Cloth, silk paintings, woodblock prints, dishes, pottery, and Kabuki masks all found their way home with me. Dean and I also bought kimonos for ourselves, mine in bright colors, his in subdued shades of brown and gray. And I acquired two beautiful Japanese Geisha dolls. How would we ever get all this stuff back to the States?

Winter arrived. The Japanese countryside under falling snow was spectacular, and the trees and houses outlined against a blanket of white were worthy of any postcard. Gregg and his friend Michael Thorpe had a great time building a snowman behind our housing unit. Gary Grau came out to help and play with them. From that time on, Gregg would often knock on Gary and Nancy's door and ask, "Can Gary come out and play?"

As Christmas came near, the Army informed us of certain safety regulations regarding the display of Christmas trees. All trees must be secured in a washtub filled with wet sand. The tree must be held upright by three wires running from the upper parts of the tree to the walls of our unit. This we did; however we had brought no tree decorations so we improvised with small decorated paper plates, ribbon, popcorn strands, Christmas cards, and strings of cranberries.

Our Christmas in Japan will always be precious in my memory. Although we missed our families and the shared joy of Christmas Eve with Dean's parents and Christmas morning with mine, there was a special spirit that we had never felt before. Here, in a foreign country that had few Christians, surrounded by the beauty of snow-covered hills leading down to the timeless Inland Sea of Japan, we felt apart from the world we had known — just us in a small world of our own. It was Coco's first exposure to a Christmas celebration. Gregg had the honor of giving her her first Christmas present.

The New Year arrived, 1956. The days passed quickly as we fell into the routine of Squadron life. I never worried about Dean's flying. The mission of the Squadron was unknown to me; had I known, I would have had great concern. Our stay at Niji Mura with its freedom, beauty, and close friends remains the highlight of my life. One by one, Base housing became available at NAF Iwakuni, and one by one our Squadron friends left Niji Mura while new arrivals and their families joined us in what was to me our private Shangri-la.

Our turn to move came in April. Unlike most of the others, we had brought our own furniture. This posed a problem, for we had already used our government paid move from the Philippines to Niji Mura. This move was ours to figure out. Dean obtained permission to use a Navy landing craft with the front end that drops down, a Japanese coxswain to pilot it, and a big Army truck with a driver. Plus, some of the Squadron men volunteered to help. The men helped us load our belongings onto the truck, after which it trundled down to the waiting landing craft and all boarded. As the landing craft disappeared over the Inland Sea, it struck me that all our earthly possessions were now floating away without us. Meanwhile, we took the train to Iwakuni and proceeded to the Base. There we saw the landing craft beach itself on the seaplane ramp and the truck roll out smoothly, then proceed to our new quarters — all was well!

We were assigned an end unit in a four-unit building. To our delight, Don and Margaret Burdg had stocked our refrigerator with food — Navy friends are the best. Next to our unit were Wells and Elizabeth Rathbun, then Gary and Nancy Grau, and on the far end were Norm and Rita Bull. Gary and Nancy and Norm and Rita would become lifelong friends of mine.

Although I missed the beauty and remoteness of Niji Mura, our new location offered a new wealth of places to see: it was just a short walk to the town of Iwakuni which lay just outside the main gate of the Base; the convenience of the Navy Exchange, Commissary, and Medical Facilities was welcome; and Dean no longer had to get up early to catch the 5:30 commuter boat for a two-and-a-half hour trip to work.

In town there were several good eating places. At Keenan's, we'd order a thick steak cooked at the table on a hibachi. It cost only 360 yen — one dollar! Gregg would order fried rice. This surprised us until we found out he ordered it only to see the cook throw the rice high in the air from the frying pan, then catch it in mid-air. We'd also go to Serra's for Italian spaghetti, where a huge wheel of cheese provided enough nibbles and pasta toppings to last for many months.

Walking through the town of Iwakuni brought many sights, sounds, and smells to our senses. Of particular note was the "Honey Bucket Man," as he was known to us Americans. Sometimes he was on foot, sometimes pushing a two-wheeled cart, but always bearing buckets of human waste — organic farming to the extreme. And evening would bring out the "Soba Man," his cart lit with paper lanterns bearing dishes of delicious noodles.

On May 5th, we witnessed the celebration of "Boys' Day" in Japan. Suspended on long poles held aloft, large paper carp fish were streamed into the wind, colorful and joyous in their undulating rhythm, much like the freedom of kite flying.

We again put our Lambretta motor scooter to good use. I would sit behind Dean, Gregg would stand between Dean's legs, and off we'd go to explore the countryside. Of particular interest was the Kentai Bridge: *The Bridge of the Brocade Sash*. Stunning in beauty of itself, spanning the river with five graceful arches, it was the essence of Japanese architecture: delicate beauty, perfection of line, elegant simplicity, and functional purpose. Wide steps covered each arch, making for an undulating up and down crossing, beneath which cormorants dived from small fishing boats into the gently flowing waters. The river banks were lined with inlaid stone. Blossoming cherry trees marked graceful paths beside which Japanese women and children in colorful native dress spread picnics on blankets and watched street performers weave their magic acts. The scene expressed the history and culture of Japan in one sweeping panorama — a shadow box of Japan at its best.

Kentai Bridge and Cherry Blossoms

Then there was the resort island of Miyajima rising from the Inland Sea not far from Iwakuni. A Shinto gateway graced the entrance to a small harbor. Fishermen would sail beneath the gateway to obtain the blessing of a good catch. Somehow, the Japanese landscape is a natural splendor of shape, balance, proportion, outline, and color. There was a "Kodak Moment" in every direction and at every step.

We journeyed to Hiroshima, *City of the Seven Rivers*, and site of the first atomic blast that brought about the surrender of Japan in World War II just ten years before. Standing by the skeletal building at "Ground Zero" brought a sense of the enormity of the destruction and power of this bomb that had propelled us into the "Atomic Age" with its mixture of hope and fear. Then visiting the Cenotaph, a shrine erected to those who died, the softness of fresh flowers and the scent of incense sending prayers to departed loved ones generated mixed feelings which swirled over us. Standing there, our feelings could not reach conclusions: horror, triumph, pity, awe, curiosity, fear, hope, and relief all vied for supremacy, but it was not that easy, for all were true. Many in future generations may brand the Hiroshima bombing as merely a cruel and monstrous act, but any such simple conclusion misses the wholeness of truth surrounding this dramatic ex-

Ground Zero in Hiroshima

pression of the tragedy of WWII. (So far only that war involved atomic weapons.) We left with the hope and prayer that such will never happen again.

Coco came over from Kure to stay with Gregg while Dean and I spent a week in Tokyo. We flew to Haneda Airport in Tokyo on a small plane (Dean said it was a de Havilland), flown by Japan Airlines. The flight was at night during thunderstorm weather. Bouncing around the sky, we landed first at Osaka where they had us all deplane and be weighed. That didn't help our peace of mind as we reboarded and launched once more into the storm.

At Tokyo we looked around on our own. Our transportation was mainly the twenty yen cab, though we feared neither we nor the driver would ever survive to the destination. The madcap taxi-drivers dashing pell-mell through the narrow streets of Tokyo with little regard for pedestrians or other vehicles still brings shudders of fear to my mind. We saw the Imperial Hotel, the Emperor's Palace, the Diet Building, a Shinto Shrine, and Japanese gardens. We shopped on the Ginza, went to the large military exchanges, and dined out at every turn. We also traveled to Kamakura to see the Diabutsu Buddha (or Great Buddha) whose serene countenance was fashioned in the eighth century.

I'll never forget going to see "The Rose Tattoo" one evening. Our seats were on a second balcony clinging to the walls high above the floor. The balcony sloped steeply downward, making us feel that we could be pitched forward into the air. Dean, six feet four inches tall, was particularly uncomfortable because Japanese theater seats are small with little leg room. But the movie began, and we settled in fairly well. Then, after an hour or so, our seats began to move and sway. The movie screen went from black and white to brown. Darkness enveloped the theater as the walls groaned and lurched. An earthquake had found us in a precarious spot. The movement made us hang on to our seats, wondering what to do as people started to bolt for the doors. Again a shock went through the structure. We stayed put, feeling that was preferable to risking being trampled by the crowd. Then, as quickly as it had started, it was over. Soon the screen lit up and the movie continued, evidently just a normal evening in the city of Tokyo.

In the last week of June, Dean came home all excited. The Squadron had just given him his certificate as PPC (Patrol Plane Commander), meaning he would now command an aircraft and crew on operational flights. He was grateful to Stretch Varner for this honor, for as Stretch's co-pilot, Dean had been given seat time and valuable training by Stretch that now bore the fruit of being trusted as a Plane Commander. After his first set of flights as PPC, he arrived home late at night and woke me up to talk and live the moment.

Gregg had new friends now. There was Patti Hansen, daughter of Bob and Jean, and "Skipper," as he was called, son of Wells and Liz Rathbun. A young Japanese girl named Sadako was our maid. She loved to watch over Gregg and would take him on a bicycle ride into Iwakuni. This delighted Gregg for they would always stop at a large enclosed cage in town, which contained a good number of snakes - Gregg's favorite critters. One of them was an albino, which fascinated Gregg. Also in the cage were small birds — meals on wings for the primary inhabitants below. Sadako also would take Gregg to a place where he could dip for goldfish to take home, for me to take care of as it turned out. Sadako quit abruptly one day announcing, "Me catchee baby" — hopefully not from a boyfriend while she was baby-sitting Gregg. Sometime in early August I also found I was pregnant. I would be due in March of 1957, several months after we would leave the Navy.

Typhoon Country

I knew them as hurricanes back in the States, but here they were called typhoons. They would sweep up across the ocean from the South China Sea and slam into Japan. I thought it one of the oddest things in my life that when a typhoon threatened, our husbands would climb aboard their airplanes and fly away to some safe place, leaving their wives behind to hunker down and face the storm's fury.

One time, as a typhoon swept ever nearer, Dean's plane was unflyable, being in the hangar for maintenance, so Dean stayed behind to ride the storm out with me. We gathered several wives in our quarters as the storm approached. Dean remained calm as the wind picked up, and the water level rose closer and closer. He remained relatively calm when the buffeting winds shook our quarters and drenching rain blew horizontally past our windows. But when he looked outside and saw an entire tree, airborne and twisting, flying away above the street, his calmness disappeared. We all clung to each other waiting for something to hit our quarters, but soon the storm abated. Outside, the streets that had flooded were now covered with mud and debris. Broken branches lay everywhere. But, amazingly, there was no damage to any of our quarters. Dean said that next time he'd much rather fly out with the Squadron aircraft.

The Day Everything Changed

All the wives knew of the danger that came with the mission of the Squadron. It was the unspoken concern on our minds. This concern burst into reality on 22 August 1956: one of our planes didn't come home. Dean's boss, LCDR (Lieutenant Commander) Milton (Mel) Hutchinson, our close friend LT (Lieutenant) Frank Flood, LTJG (Lieutenant Junior Grade) James B. Deane (a newcomer to the Squadron), LCDR James W. "Pappy" Ponsford, and a crew of twelve more men were shot down by Chinese fighters over the Bay of Shanghai. The wives rallied around their grief-stricken friends while waiting for news. Were they all killed? Was there any hope for survivors? Several days passed with no encouraging news. Then came word that the Seventh Fleet, in their search of the area, had found the bodies of two men and some wreckage, all returned to NAF Iwakuni. Dean was summoned to help identify the two who had been found, LCDR "Pappy" Ponsford and AT1 (Aviation Electronics Technician First Class Albert P. Mattin. Then we learned that the bodies of two enlisted men, AT3 (Aviation Electronics Technician Third Class) Jack A. Curtis and AT1 William Frederick, had washed ashore in Shanghai Bay and were being returned via Hong Kong. And that was that. The enchantment of our stay in Japan disappeared. The world seemed stark and cold.

Dottie Flood and I had formed a close friendship. We had been scheduled to go to Hong Kong together with Dean and Frank. Admiral Anderson, Commander of Taiwan Patrol Forces and Commander of Fleet Air Wing One, was to visit the Squadron at Iwakuni, then go to Hong Kong in his ship. Dean and Frank had arranged to ride with him, while Dottie and I would fly there on CAT (Civil Air Transport) Airlines, which had been founded by General Chennault of Flying Tiger fame. But a typhoon sprang up over the South China Sea, causing the Admiral to cancel his trip. Then Frank received word that his dad had died, but Frank elected not to return to the States for the funeral, saying he and his dad had said their goodbyes before Frank left for the Far East. At the Squadron, Hutchinson's crew was short a copilot who was scheduled to leave the next day for the States. Frank volunteered to fill the vacancy for that mission. Somehow it seemed that Frank should have been somewhere else than on that doomed flight.

Dottie returned to her homeland, Canada, and the other wives left to find a new life for themselves. I wonder what life holds for them. The fate of Mel Hutchinson, Frank Flood, James Deane, and nine crewmen was never determined. The chances of their surviving were very slim, but tell that to a Navy Wife. How can hope not linger, no matter what the odds?

Homeward Bound

Dean and I never intended to make the Navy a career. All along we looked forward to returning to Salt Lake City and finding our life there - to rejoin family and friends — to pick up the life we left behind. Orders arrived for our transportation home. We were excited in the knowledge that we would soon see our parents again and see Dean's sister Elaine, who married a fine young man, Kenneth Judd, while we were overseas.

Just as we had needed shots before we traveled across the Pacific on this tour of duty, we found we needed some similar attention before going home. It seems the vegetables in Japan carry various larvae of intestinal worms, including the tapeworm. That is why we had been cautioned to soak all vegetables in a mild chlorine solution before eating them. And so we reported to sickbay to be dewormed. The medicine was heavy and sweet, but we all managed to down it. I took their word for it that it did the job.

The Squadron threw a big going-away party for us. On the wall was a huge paper banner saying "Sayonara" above a ship heading for the States, from which Dean and I were flying kites while Gregg was flying his kite from a small boat towed behind. Our destination was aptly shown as "Salt Lake."

My feelings were in a jumble the next morning. I was excited about returning home to bear my second child and have Gregg get to know his grandparents, but sadness also swept over me. I was leaving dear friends that I might never see again. Life had brought our paths together, and now our paths were diverging. The future for Dean, Gregg, and me seemed uncertain. It was time to take a deep breath and have faith that all would be well.

At the airfield we boarded a Marine "Flying Boxcar" for the trip to Tokyo where we boarded the USNS Morton, the smallest military transport of the fleet. The ship was carrying some 4000 Marines and a few Navy personnel and their dependents. We were assigned a small but comfortable stateroom. We departed Tokyo harbor late in the day. As night fell, we encountered rough seas. Dean was assigned to stand a fire and security watch. It was his first time on a long sea voyage. I kidded him that I had logged far more sea-time than had he.

The ship's motion did not agree with my pregnancy. Nausea overcame me, so I stayed in our stateroom. Gregg enjoyed the rolling motion. He would line up his little cars along one bulkhead, then watch them roll to the other bulkhead and return with every roll of the ship. I tried not to watch. We had to "dress" for dinner, including Gregg's wearing a white shirt and bow tie. The stewards invited Gregg to help ring a hand-held gong that sounded out the calls to dinner. The food wasn't good at all. We joked that anything worth eating had been devoured on the crossing to Japan, leaving leftovers for the return trip. I became nauseated during one dinner to the point that I left for our stateroom. Gregg was very upset and admonished me not to do that again.

The days at sea became routine, leaving much time for thought. I remembered how uncertain I was about the future when the USNS Barrett set sail for the Philippines. Now that the excitement and rush of leaving the Squadron and Japan had passed, various emotions swept me. I realized what good friends I had left behind, although I knew that sooner or later all would leave on orders to new and unfamiliar destinations. I also became aware how much I had enjoyed being a Navy wife. I had been raised to be dependent on my mother. She loved me dearly, but had a rather domineering personality and had made most of my decisions as I grew up. She lived her life through me, an only child. Now, after five years as a Navy wife, I had become independent and confident in my ability to take care of myself, my husband and my children, and I needed no one to make any decisions for me. I wondered how my "new self" would relate to my mother. And I enjoyed the new experiences that each set of orders brought. After two years in one place, I was always ready to move. Would I be satisfied living in one place for the rest of my life? And here we were, expecting another child, yet Dean had no job prospects. How would this all work out? Do we even want the life we left behind? The rolling sea held no answers for me.

USNS *General Morton* passing under the Golden Gate Bridge

Each day as the ship drew

nearer to Stateside, Japan grew further away in memory as well as in miles. The gentle movement of the ship and the soothing trembling of the deck beneath my feet made for ease of reflection and thought. I became aware that a precious part of our life had closed - did it really happen or was it all just a dream? Then memories of Niji Mura flooded my mind. The mystique of that isolated haven across the waters of the Inland Sea will forever remain close to me. The terraced hillsides, green rice fields, cherry blossoms, rice-paper lanterns, and sparkling waters of the Inland Sea bearing fishermen in their boats have become etched in my memory. Again I picture little Gregg chasing across our yard in Iwakuni, leaping into the air as he tried to catch butterflies with a butterfly net we had given him. I remembered Gregg and his friend, Patti Hansen, hugging each other as they said goodbye. And tears of grief and anguish came again for Dottie Flood and the others whose husbands were lost to them and to the world. But this I did know: our life had not been wasted in the Navy — we lived — we served — we grew in life skills and abilities — we made lifelong friends, and we knew what it was like to lose close friends — we went to places we never dreamed of going to — and we had an adventure that few are fortunate enough to experience. I recognized a growing feeling within that I would look at the past two years as the best years of my life. How blessed can one be?

Epilogue

LaVon and Dean returned to their families in Salt Lake City, where she gave birth to another son, Brad, on 22 March 1957. Dean gained employment with the Civil Aeronautics Administration at the Salt Lake Air Traffic Control Center. A year later, they returned to active duty in the Navy, in which Dean completed a twenty-three year career. After a second career in test and evaluation of military systems, he retired and they built a country home in the rolling hills of Virginia near their son Gregg, his wife Michelle, and their two children, Erin and Joshua. Brad and his wife Julie made California their home with three children, Israel, Hillary, and Tiffany.

LaVon died on 11 January 2006 and was buried in the beauty of Utah beside the magnificent mountains from which she came, the land of her Mormon heritage. She is remembered by her family and all who loved her. Next to her roles as a daughter of Heavenly Father and mother of her sons, was her love and pride of being a "Navy Wife."

The remains of the other crewmen lost over the Bay of Shanghai were never found.

Written by LaVon's husband, Dean B. North.

Nancy Lee Grau

LIFE BEGINS AT 23
My Rather Short Life as a Navy Wife

At last THE phone call came, the one that told me we actually would be having a life together. We'd been married just before Christmas and spent an altogether too brief time together. Then we separated again with one of those gut-wrenching departures, which, I was learning, would become customary and even more frequent. My new husband Gary completed his training in multi-engine planes at NAS (Naval Air Station), Hutchinson, Kansas. I resumed my first year of teaching in Des Moines, Iowa. It hadn't been the most ideal way to begin a marriage but probably wasn't really all that uncommon in 1955, near the end of the Korean War.

The call told me he and three other men were to be assigned to a squadron called VW-3 Det A (Detachment Able) "Somewhere in the Pacific," following a month of leave. I would learn those names and meet the men eventually. At this time Gary knew only one of them. That was bachelor Karle Naggs who had been in Pre-flight a class or two ahead of him in Pensacola, Florida. The other two men were married. Frank Flood and his wife Dottie would become good friends of ours. Milt Hutchinson was also married, but our acquaintance with them seemed only to be when we greeted each other briefly at the occasional Squadron parties they attended.

Not during our phone call but much later, Gary told me of his crushing dismay at the assignment. He had requested (1) a specific squadron in Bangor, Maine; (2) any squadron in Bangor, Maine; or (3) any squadron on the East Coast. He'd scrupulously hunted the posted assignment list for his name and finally found it sandwiched between those of men going into helicopters and those being sent to "poopy bags" (balloons). When he saw his assignment, he went to a master book and discovered that the weather surveillance squadron VW-3 was headquartered in Guam. With an increasingly sinking heart, he feared that any detachment of a squadron had to be on an even smaller island, like Pago Pago. He wondered, "What on earth did I do wrong?" Immediately he went to talk with his instructor and was surprised to see a twinkling eye and rather knowing smile as the instructor said, "That's my old Squadron." He was Dick Renner and once had been in the Squadron, possibly even before it was designated VW-3 Det Able. Without giving Gary any specific details, he told him he would enjoy the experience of being in that squadron. He should be very aware that soon he would be the subject of rather intense investigation to give him a high security clearance. Again, without receiving any specific information, Gary somehow was not surprised when he learned the Squadron detachment was in the Philippine Islands. Mr. Renner, though, did tell him that he and the other three men ordered to the Squadron were the first ones who were not volunteers. (This was later disproved.) From this, Gary surmised there might be some danger — and excitement — involved.

The good news for me was that I'd be able to join him after he was overseas. Also, he'd have a month's leave before going there. Operating under the idea we'd be together on Guam, but more excited about the immediacy of his leave, right away I went to my school's principal to request a month's leave of absence. She was most cooperative, saying there would be no problem finding a substitute teacher. School district policy recognized that unhappy teachers were not very effective in the classroom. When she learned Gary might be going to Guam, she picked up the phone and made an appointment for me to meet two new couples in the school district who had recently returned from there. They invited me for dinner, were very pleasant and quite enthusiastic about living there, and gave all sorts of do's and don'ts advice. Eager to learn every last morsel of information about my potential future home, when someone referred to an area as "out in the boondocks," I pounced on this exotic new term.

"Is that a Guamanian word?" I inquired.

After a sudden hush while each of the four tactfully and carefully avoided each other's eyes, one nodded and said, "Yes."

We spent part of the leave at his parents' home in Chicago and ended up with my folks in Fort Dodge, Iowa. Dick Mester, one of Gary's friends with whom he had received his wings, lived in a town north of there and came to Fort Dodge to pick him up. The two of them then would drive to San Diego where they were to undergo more schooling. After this new parting, I returned to complete the school year in Des Moines. With several security checks, Gary's parents learned about from their neighbors, I was surprised never to hear any about me. Uncle Sam must have had great confidence that the men would not talk to their wives about their activities or that we wives would spill the beans should they slip up some time.

Gary departed for NAS Sangley Point at Cavite City, Republic of the Philippines. Over July 4th he came back to the States to pick up a P4M *Mercator* plane for the Squadron, giving us another few hours together with his parents in Chicago. By this time we knew when I'd be joining him, and I'd begun taking whatever steps the Navy told me were necessary before my trip.

I don't recall any problems completing the paperwork promptly, and my passport arrived without memorable trouble. The shots were an interesting experience, most of them without incident. I went through the prescribed series for several weekends in Fort Dodge starting well before school was out. But one of the last ones was particularly virulent, not the shot itself, but the results. Fortunately this happened after the school year ended. The summer weather was glorious — but warm — so I decided to walk the two miles home. The shot (cholera, typhoid, or typhus?) had been in the hip, and apparently the heat and action of walking did something to hasten or strengthen its effect. Within an hour of arriving at home I had a raging fever. Since this was not long before my scheduled departure, I didn't want anyone to know how ill I was for fear it might keep me from the trip, even though quite convinced death was right outside the door. My mother had one of her infamous three-day migraines, so I had promised to take over cooking duties. I was so weak and dizzy that crawling on hands and knees was the only means to the kitchen. Somehow dinner was prepared and eaten, quickly followed by retiring to bed. The fever didn't seem to last long but it certainly was intense!

Unpleasant as this experience was for me, it probably didn't compare with the one Gary had. While he was going to the San Diego school, several senior officers came up to him, gave him a conspiratorial grin, and said, "I know what you're going to be doing." (Mr. Renner had told him this would happen, but they really didn't know anything.) Despite all this interest in what Gary was or wasn't going to be doing, no one seemed to have thought to see that he got his shots. Two or three days before he was to depart San Diego and orders were being cut, it suddenly occurred to them he hadn't had any! So they sent him through double lines where he got all the originals in both arms all at one time. He left and arrived in San Francisco on Friday afternoon to get his transportation arranged. There they gave him all his required booster shots. He was somewhat soothed with the idea that he'd have at least the weekend and perhaps even longer to be in San Francisco. He could start feeling better and have excellent meals as well as a little relaxation. Out he went for the first wonderful dinner, then headed back to his room for the night, only to find a message that he'd be leaving first thing in the morning. With his extremely sore arms, it was a wonder he could carry the two heavy B4 bags that held all his worldly possessions.

So now I faced the dichotomy of leaving my family, versus the interminable three weeks of anticipation before leaving to join my husband. I was on my way to a foreign land, shots, luggage, and all. I was going to travel halfway around the world to live for the rest of my life with a man I really didn't know all that well, nor he me. However, we both had faith in each other and with youth's confidence were sure we'd made the right decision in choosing each other. Despite our spotty courtship, that faith probably was the only way the relationship could have survived. In retrospect, this was the first step in what would be a two-year honeymoon.

My parents drove me to the bus depot for the last time, the first of the three means of transportation to be taken that day, followed by another one — a ship — a few days later. My mother said, "We don't want you to leave, but we realize your place is with your husband now." For some reason I didn't cry; I think I was just too excited at the prospect of my new life. It wasn't until somewhere in the day's transportation marathon that it struck me hard I wouldn't be seeing them for two whole years. We'd been a loving family, but I'd been having another life and making nearly all my own decisions without parental input for four college years and beyond. My father did cry, perhaps he had the same fears about me as I did about Gary, and that when we parted we might never see each other again. I looked at my brother and realized he'd be a different person when we next saw each other. At age six I'd starting asking for him and had to wait three years. We'd just celebrated his fourteenth birthday. He'd been my doll, my charge, my student, my friend. Still my little brother, he'd soon be a young man with a deeper voice, heavier eyebrows, and needing to shave. Taking the bus from Fort Dodge to Des Moines was the last customary thing I'd do.

In Des Moines there was time for a few hours' layover at my uncle and aunt's, Boyd and Leola Porter. While I was teaching, the weekends I didn't go home had been spent at their house. That was the last time I saw Boyd's mother, my step-grandmother, whom I hadn't seen since I was a little girl. When I got on the train to Omaha from where I was to fly to San Francisco, I realized that I really was leaving *everything* familiar to me to enter a brand new world. Apprehensive about my first airline flight, I took Dramamine which, coupled with the plane's very delayed departure, made me sick. I'd tried to sleep on a wooden bench in the interim between first and second boarding's of the plane, worried because the "something" mechanically wrong with it that caused the second boarding didn't set my mind at ease. The sleep effort didn't work, and I was exhausted. By this time it was perhaps 4:00 or 5:00 a.m., after a very emotional day of good-byes.

What on earth was wrong? When I arrived at San Francisco it was freezing! Here I was, in sunny California and going to the warm-climate Philippines. I had only one long-sleeved outfit, the linen suit I was wearing. The people I saw around me this July day were in wool suits with winter coats! No one I knew had ever been in San Francisco to let me know what July is like there. I was so tired I could scarcely appreciate the new scenery around me. Apparently with my crumpled suit and haggard appearance, I looked like a lost waif and/or pauper. A kind-hearted taxi driver directed me to a shuttle bus, saying it would cost me too much to take a cab the fifteen miles into the city. At Gary's instigation, I fearfully had written total strangers, his Uncle Hugo and Aunt Jan Grau in Stockton, California, asking if I could stay with them for the introductory days leading up to departure. When I got to the bus terminal, I phoned Uncle Hugo my arrival time in Stockton, and he met me at the depot there.

Hugo and Jan and their two sons, Dana eight, and Gregg six, showed me a wonderful time for those few days, with lots of swimming in friends' pools, etc. I had one or two other bus trips back and forth from Stockton to San Francisco to take care of necessary governmental matters before sailing, and then, sensing my apprehensions and timid plea, Hugo drove me into the city the final day.

One of those extra bus trips had been to hand over specific papers concerning my departure, and this brought about my most embarrassing incident. It was in regard to the medical permission slip. Our government had so many multiple-copy forms to fill out for the most insignificant things that I simply could not believe a tiny single slip with the words, "I have examined ______ and found him/her physically fit," could possibly qualify for my extended stay in the Orient. Convinced someone had slipped up in compiling my packet, I'd written for another medical form — and received the same one. Still unconvinced, I had my physician write out a detailed explanation of my healthy condition. Since he described my entire body in great detail, it was excruciating to have to stand in front of a young corpsman for a longer time than anyone else, in that huge packed room. He slowly read each word, from time to time glancing at the specific parts of said body as if to verify each item.

LaVon North, the wife of one of Gary's friends (Dean North) in the Squadron (now designated VQ-1 — Electronic Countermeasures Squadron One at an impressive commissioning ceremony June 1, 1955), was to be traveling on the same ship. We had made arrangements to meet in San Francisco during one of those extra bus trips, probably at the hotel where she, son Gregg, and her parents were to stay. Her parents had driven them from Salt Lake City

LaVon took me under her wing for the first of many times, lending me a long-sleeved sweater for the hours I was in San Francisco. She also began educating me about Navy life. Our ship was to be the USNS *Barrett*. Since she had three-year old Gregg, she was assigned to a cabin — a very small one — with an older lady and her child. Somehow I was lucky and got in with three other wives about my age, two of whom were pregnant. Two of them, June and Perky, were physicians' wives. Unfortunately one of the pregnant wives seemed to prefer only her own company. Sad. In response to one of the other's saying, "I'm just positive about ____," she made her only memorable comment. From her self-imposed isolated corner came a contemptuous, "Only fools are positive!" Hardly endearing! We subsequently left her to the company she liked best. Her withdrawal from socialization was so complete, I wondered how on earth she got pregnant. Perky, the other pregnant one, was very warm and friendly as was June, our fourth cabin mate. We heard from Perky for a few years after our paths separated. The three of us had a lot of fun together, just as if we were college friends, and we did get to meet each other's spouses at a get-together a while after our arrival in Cavite City. Our cabin was comparatively luxurious and spacious with our own reasonably sized bathroom, just below the one assigned to the family of the new Commanding Officer of NAS Sangley Point.

Reading sea stories had been a major interest during my growing-up years. Now elated that my great adventure had begun, I spent the first hour outside on deck, watching us go under the Golden Gate Bridge, feeling the salt wind blow in my face, and picturing every one of those romantic stories. At last I was really on my way to the new life! I had liked to amuse myself and others with the obviously absurd comparison that my limited experience on rowboats and an occasional sailboat and speedboat would relieve me from seasickness on a ship, stoutly maintaining that was just a psychosomatic malady. As the ocean got rougher and rougher, I reveled in the stimulating movement and wild wind blowing in my face. Very soon it was the first dinner seating at 4:20. Still euphoric, I joyfully went inside and started down the ladder (Navy for stairs) to the dining room. My attitude rapidly changed when the aroma from the food rose to greet me. The food in my stomach rose as well, and I barely made it to my bunk. There I remained horizontal for two days, not even able to raise my head from the pillow without becoming severely nauseated. My cabin mates became concerned about dehydration. At my request they finally brought a large icy Coca Cola. I braced the drink between mattress and bunk's side frame so with the aid of a flexible straw it could be sipped from time to time. The next day I was just fine. When we ran into some rough seas after departing from Guam later in the journey, this time without getting sick, I again rejoiced in the exhilarating rise and fall of each wave that raised and lowered me off the bunk.

The ship was quite nice. Considering that later I heard several of the Squadron wives talk about the ships they were on, it must have been stupendous! Their recurring complaints seemed to be about constant paint chipping outside their cramped cabins, bad food, etc. Our voyage seemed more like a vacation cruise.

Our routine was not terribly demanding. We sunbathed before 9:00 a.m. because the sun then became much too hot. The sundecks became centers of our social life, and lots of the time we sat under a canopy, watching Gregg and other children playing under another canopy. Visiting with other passengers seemed to be the major activity, and LaVon spent as much time with us three girls as possible. Unfortunately her roommate was not at all hospitable, possibly in part because their cabin was so small. I don't think it even had a place to sit down, just two sets of bunks. It certainly was way too crowded with two small children in it. We felt bad for LaVon and would have loved to exchange her for our unfriendly cabin mate — we thought she and LaVon's roommate deserved each other!

The first few days after I became mobile there were lifeboat drills, and after a couple of them the ship's crew informed us that if there was another it would be for real. Despite their assurance, there was another, still just a drill, but a couple of women nearly went into hysterics because they had once been on ships which were torpedoed not far out of Seattle, either just before or just after World War II began.

Most of the movies scheduled for our entertainment were disappointing selections, so we just skipped them until the last day when we were treated to the riveting "Rear Window."

Besides dependents, this was a troop ship. We wondered where the troops were. There was a large empty deck at the same level as our cabin. Since it was always deserted, we felt no qualms about leaving the portholes uncovered when changing clothes. About the fifth day one of us glanced up, noticing the light had darkened. The portholes were plastered with faces of young men staring in at us! We now knew where the troops were. We altered our dressing procedure.

LaVon was very helpful in educating this very naïve girl about Navy life. I'd never been out of the Midwest and knew absolutely nothing about military life, save what we kids played for several years and what we saw in the movies. The pleasant NAS Sangley Point Commanding Officer's wife began to get very friendly to me. LaVon warned me that this sometimes happened with young wives and to watch out, she wanted something. Sure enough, she began asking me to baby-sit and do various other little favors for her, not anything very demanding. It really didn't bother me much except to wonder how I'd handle it if she continued to do it once we were off the ship. One day on the sun deck, the Commanding Officer himself asked me about Gary and what he did at NAS Sangley Point. Since I didn't really know, and couldn't have answered even if I did know because Gary was in a very secret Squadron obviously doing dangerous work, I even wondered if this was some sort of a security trap. I didn't know if I could even say what his Squadron was. I replied I thought he was the Navigation Officer, which he was currently within the Squadron. The Commanding Officer attempted to get to know others of us who were headed to Cavite and one day asked LaVon what her husband did. She was not quite so reluctant and responded, "VQ-1, as is Nancy's husband." It got very still. His wife no longer asked me for favors. VQ-1 was an entity by itself with its chain of command through only the Commander of Naval Forces in the Pacific directly to Chief of Naval Operations (CNO). Even though VQ-1 was stationed there, NAS Sangley Point's Commanding Officer had absolutely no jurisdiction over it. When I later saw them at the swimming pool and the Officers' Club, they were always very pleasant, but that was as far as the relationship went.

While our ship got nearer and nearer to the tropics, the sunsets became progressively more spectacular. One particularly magnificent one that lingers in memory was the entire horizon encircled with purple clouds, every one of them edged with gold. Would that I really could describe this splendor!

We went thirteen days without seeing any sign of life, save for a few sea gulls during three or four days out when we passed somewhere between, but not in sight of, Hawaii and Midway. I began to think like Noah. "What if something has happened to every land mass and the whole rest of the world has been destroyed? We are so self-sufficient; will we just keep sailing until the supplies run out?"

Resentful and rebellious crewmembers mutinied against Captain Bligh. They cast him out from the HMS *Bounty* with a few loyal men in an open vessel hardly larger than a rowboat to face almost certain death as they miraculously traveled thousands of miles in forty-five days. Captain Bligh certainly gained my respect and admiration for his survival. For us, a tiny speck appeared on the horizon the morning of the thirteenth day. I stood on deck for four hours watching as the island of Rota came and went. It gave me neither an anticipated tan nor sunburn but instead some sort of sun poisoning which covered my skin with pinpoint red dots. However, I had seen land, and it dispelled that rather disoriented feeling of isolation. And then we arrived at Guam which excited my curiosity since I'd thought it was where I was to live for two years.

The island of Guam is about twelve miles wide by thirty miles long. Some of us took a bus tour of the northern part of the island, driven by a brilliant seaman who got lost and three times showed us Anderson Air Force Base with its huge B-36's. We had a short shopping spree in Agana, Guam's capital city. Then we were glad to return to our home-away-from-home, thankful not to be stationed on Guam, and eager to complete the home stretch to see our husbands.

But one of my cabin mates was missing. June had gone to see friends stationed on Guam. When they brought her back to the boarding area after dark, the guard on duty said he'd see that she got safely to the ship. Instead he drove her into the southern part of the island, then off-limits, riddled with World War II Japanese mines and containing a deserted former Japanese airfield. She talked and talked and talked, and that's what he said was all he wanted. Eventually she talked him into bringing her back about midnight. She was nearly hysterical and said the only thing she could think of was that she'd come all that way only to have something bad happen to her. I don't think she reported him. Probably not, because he would have been in deep, deep trouble for deserting his post, and nothing really happened to her — except stark terror.

About 8:00 the evening of the fifteenth day at sea, we entered the Philippine Islands' San Bernardino Straits between Luzon and some smaller islands, traveling through them until early the next afternoon. The scenery was exotic. Already lush colors, principally various shades of green, intensified, as they often do just after sunset and before dark. The high pointed hills were layered with terraces all the way down to the water, except in some places where there was a relatively flat shelf with a few isolated Filipino homes called nipa huts. It was a welcome change from the sea's beautiful but endless blue and a feast for the eyes until darkness took over.

The next morning LaVon and I were more than excited. Dean had told her that he and fellow Squadron mate Stretch Varner might fly over the ship as it approached Manila Bay. Sure enough, a multi-engine plane actually flew right alongside our upper deck where we waited in eager anticipation! We could see inside the cockpit, and both pilot and co-pilot appeared to be tall (both Dean and Stretch were about maximum pilot height). With great glee we told everyone it was from "our" Squadron and contained LaVon's husband. Fortunately most of the people on the ship were going on to Subic Bay, so we didn't see any of them later to have to explain that it wasn't one of our planes nor anybody we knew.

The entrance to Manila Bay is guarded by Corregidor, and that brooding historic island made me feel a shiver of dread. We, the children of the Depression, were also the children of the Second World War and awakened by that war to the very real presence of the rest of the world. There probably is not an American alive today who was then above five years of age and does not remember precisely his or her activity and locality on December 7, 1941. Glued to radio news broadcasts, we were the first to have war in our living rooms. We did all the things which were our patriotic duty, like making balls of the tinfoil then encasing Hershey candy bars (when we could get them) and sticks of chewing gum, gathering milkweed pods for Navy life preservers, trying not to grow too quickly out of our rationed shoes, and clutching carefully hoarded ten pennies tied in the corner of a handkerchief as we proudly marched to the Post Office to buy a U.S. Savings Stamp. This we kept in a special book to be traded for an $18.75 U.S. Savings Bond when there were enough books; if we kept it ten years, we could trade it for $25, our first lesson in investment. Gas rationing in our town usually was not a big problem — cars were customarily taken out only on weekends or for special events. We were used to walking everywhere. And maintaining top automobile speed at 35 mph did not seem a hardship. The thought of 60 mph - an actual mile a minute! — seemed an impossible achievement. Playing "War" or "Soldier" easily replaced "Cops and Robbers" or "Cowboy." Our favorite tree with a vine hanging from it changed from Tarzan's African locale to part of the South Pacific jungle. So did an adjoining large, dense grove of lilac bushes crisscrossed with various pathways. Some of us even were Jr. Commandoes, marching up and down the sidewalks for Saturday "drills." We learned a different geography from that presented in our schoolbooks. My study of the war as it happened was in ***Life Magazine***, subscribed to by the shoe store my father managed so that the owner could serve in the Navy. Corregidor and the Bataan Death March made a huge impression.

Ruins of Corregidor. US Navy photo

So did seeing them. Shallow Manila Bay was an incredibly stunning sight: I'd not been prepared for the shock of seeing what seemed to me to be hundreds of abandoned, half-sunken Japanese ships. The answer to my later letter of inquiry to the Navy said that at one time there were about 750 of them, largely gone by our 1955 entrance. A harbor pilot came out to help us thread our way through the maze. A cloudy, windless day, it took us two or three quiet hours of stifling, slow-moving sultriness to get through to our mooring site. Some of the ships seemed nearly close enough to touch. Some had vegetation growing on them. They were very silent in their reminder of what had filled the news of our young lives. The thrumming engines which had so dependably carried us thousands of miles, and even worked extra hard to get us here a day early, were now scarcely audible. As we approached the pier at Manila, the pilot's boat came out to pick him up. LaVon and I had proudly showed each other photographs of our handsome husbands. Standing on the Captain's deck, each of us looked down and turned to the other saying, "Isn't that your husband down there?" Indeed each was, standing side by side at the edge of the pilot's boat, and we had quite a laugh over picking out the other's mate instead of our own. To make sure we were seen, I frantically waved a stuffed toy Gregg had been holding. Gary interpreted it that I was telling him I was pregnant! Wrong.

After joyfully greeting our loved ones and exchanging introductions of each others' spouses on the dock in Manila, we got onto a much smaller boat. This would take us across the bay to the Naval Air Station at Sangley Point. After dinner at the Officers' Club, we headed off to our individual homes. Gary and Dean had borrowed a car for the occasion. The house Gary had rented for us was at 256 Lopez Jaena Boulevard, Cavite City, and was closer to the Base than their apartment compound, so they dropped us off and went on to their own new home.

We unlocked the front gate and gazed at our very first home. We'd been married eight months and never lived together, other than a few honeymoon days in Gary's sister's apartment. She'd been married six months before we were and as a wedding present to us, she and her husband temporarily moved out of their dollhouse-sized apartment behind Gary's parents'. Our home now-to-be was a single unit house owned by Mr. Abarro, a member of the Philippine National Bureau of Investigation. On the other side of a fairly high wire fence along the right side of

Gary and Nancy's first home in Cavite

our walkway were three or four native houses, one behind the other, nipa huts on stilts high enough so the houses could be walked under. Our large front yard had a grassy path to the house, bordered on its left side with purple and white San Vicente flowers. These pretty flowers were also along the fence at the far left of the yard. Our walkway was close to the right hand fence. Between the fence and walkway were exotic-looking food-bearing plants, some in the ground and some with gourd-like fruits hanging from long vines running overhead. We also had several banana trees and pineapple plants. Standing on the windowsill of the first house was a small boy, completely covered with sores which we assumed were yaws. Across the left fence was just one house, about even with ours with each back from the street quite a ways. Both it and ours were raised slightly, but our mini-stilts were hidden by lattice work while the other had naked posts.

Although Gary already had rented the house for a month, he'd never stayed in it. He'd hired a house boy, Alphonso, and a maid, Loretta, to make sure it was in living order. Loretta came in five days a week for her $15 monthly salary, Alphonso three for his $7. Every day she did our wash by hand, including sheets and towels, in a low, flat, round galvanized metal pan with a serrated edge which sat on the ground. We had a clothesline for drying. Some Filipino people did not, so their wet clothes were scrunched together on low benches covered with pieces of a favored building material, corrugated tin, or lacking that, on the ground. Each day he came, Alphonso polished all our shoes, whether they'd been worn or not, and painstakingly cut the grass by hand with scissors. Loretta was a shy, quiet girl who spoke almost no English and scrupulously attended to her work. Alphonso was bright, ambitious, and personable. After about a month he left us to work at the Base Ship's Store, surely a step up. We were happy for him but sorry to lose him, especially when we got his replacement, Ignazio, who displayed none of the same desirable attributes. He spent a great deal of time resting — from what? He was a study in slow motion.

Alphonso, the House Boy

This first home as a married couple was very colorful. The exterior was two shades of blue, and there was a small front porch floored with red tile. We saw an amazing spectrum when we entered the house. The living room walls were pink with trim a very hot pink and some other major but no longer remembered color. A built-in bar between two doorways was red with a brown counter-top. To allow good circulation, all the interior walls were latticed at their open tops in even another color, perhaps yellow. Fortunately the outer wall, also open at the top, had screens over the opening. All the walls, inner or outer, reminded me of the summer lake cottages of my youth because they were only a board thick. We had three bedrooms. The yellow one was at the front of the house and probably had been used as an office when Mr. Abarro lived there because it had a separate entrance off the little front porch, its other access only through the middle bedroom. That middle one was pink, and the back one, the largest, was blue. The second night we discovered that each bedroom had fluorescent lights matching its walls' colors. Most of our outer walls were filled with sliding windows over screened openings. When closed (like during a typhoon), the otherwise never-closed screened windows permitted some light but no visibility. They were segmented into small frames of wood about two inches square filled with a translucent material called capiz shells. The living room furniture consisted of a couple of chairs and a love-seat. Their brightly flowered cushions were framed by substantial bamboo or rattan. Each morning we discovered little piles of sawdust on the floor under the upright front curves of the wood. At first we were just curious about those piles, then realized termites had to have found the wood a tasty treat. However, as the time we stayed there lengthened, we became increasingly careful about sitting down, wondering how long it would take until the chairs collapsed. We also wondered if there were termites in the wood of the house, particularly the posts holding it up.

About the same time he rented the house, Gary ordered a bed for us from the Ship's Store. It hadn't arrived yet. In amazement, we looked at the existing bed, located in the pink middle bedroom, the most private one because it had the least outside exposure. The bed aroused our curiosity. First, we could see it was considerably shorter than either of us and was not too much wider than a twin bed. Then, as we lay down, we quickly divined that the two-to-three inch thick mattress was probably stuffed with coconut husks. So were the chairs in the living room and, believe me, they would never be a comfortable substitute for a real mattress or cushions! We struggled to figure out how we could arrange our bodies to fit either around or on top of those mattress husks, all while our legs were hanging over the end. It was especially hard to do while we were laughing to tears! About the time we'd sort of placed ourselves in some reasonable position, Gary remembered that our hydration system was a tank on the roof that filled at night when water pressure elsewhere in Cavite was less used. To turn it on and off required manipulation of a complex set of valves several feet above the ground. They were located around the corner from a back porch on the outside of the house. This necessitated planting one foot precariously on an extremely narrow shelf running around the back at the same level as the back porch and house floors. Only when your foot was as secure as it could be would you dare to lean over to reach the valves, all the time trying to remember which valves turned right or left and which ones controlled the flow into the house. Mr. Abarro had the system hooked up so that neighbors could use it, too, hence the reason for the multiple valves. This eccentric hydrating procedure was a sort of security system for us, as evidenced by the Filipino neighbor lady who sat on our porch every time we left, staying there until we unlocked the front gate. Then she slipped away to her home. We never met her.

Gary took care of it all quite nicely and came back to bed. In a surprisingly short period of time, we heard water splashing, which we assumed meant the 50-gallon tank was full. So he repeated the precarious process, reversing the valve directions, and returned to bed. In another brief time, we again heard water splashing. He said, "Hmm, I must have turned the wrong valves," so again repeated the previous step. Twice more this happened, to which he finally disgustedly remarked, "The hell with it, let it go!" The next morning as we ate breakfast, amazingly we heard the same splashing. This led to the discovery that in the house closest to our dining table, one corner apparently was used as a bathroom toilet and was cleansed by throwing a bucket of water on it after the necessary task had taken place. It was not an appetizing concept.

Another interesting discovery was in our bathroom that first morning. When Gary turned on the light to shave and looked in the mirror, he'd lost his lips! The overhead fluorescent light was green and it washed out all the natural skin pigment, making him feel as ill as he looked. We were able to pull a floor lamp in from the living room for him to finish the job. We quickly learned that for a warm water shower we had to wait until afternoon so the sun could heat the water in the rooftop tank. The shower was located in the middle of the bathroom and we had to walk past it to the toilet stool. But we were grateful to have one.

Fortunately our bed came soon, and we moved into the blue bedroom in the back where corner windows enabled more moving air to circulate, if any was moving. The front yellow bedroom now housed our shoes, all neatly arranged by Alphonso along one wall. Our former middle pink bedroom became the dressing room. It had two "hot lockers," basically two wooden armoires each fitted with a low-watt light bulb kept on at all times. Supposedly this would help keep damp and mildew from anything made of cloth. Before coming on my trip, I'd bought cotton underwear, having been told not to bring any made of nylon because the ants liked to eat garments made from it. In the dressing room we also had a rattan clothes hamper. The termites seemed to enjoy it even more voraciously than the living room furniture but fortunately they didn't seem to care for cloth.

I had quite an extensive shoe collection. My father sold shoes, and several pairs of mine were special orders because my feet are very narrow. It was a serious tactical error, however, to bring two pairs of very strappy high-heeled sandals, assuming they would be cooler than shoes that enclosed feet. I hadn't thought about how my legs ached from "growing pains" as a child and how my feet and legs swelled in hot and humid weather. I never realized until much later in life that the condition might have been caused in part by my compulsory daily consumption of eight ounces of carbonated beverage and favored salty snack du jour. Heaven forbid I should not wear those shoes to any dress-up party! One particularly memorable evening was when the straps were so imbedded in my swollen feet, Gary had to help pry them off when we arrived at home. And it was through tooth-clenching determination that I learned to wear new-to-me black leather flip-flop Oriental slip-ons, even though the strap between big and first small toe made my shin bones feel like there was a wedge splitting them apart.

Back to the house, there were no upper kitchen cupboards, just open shelves for the dishes Gary had carefully selected and I dearly loved. Our stove was a two-burner hotplate. Gary had borrowed a waffle iron from Squadron mates Bob and Jean Hansen so we could use its flat plates for additional cooking. We'd had cold cereal for breakfast that first morning, so didn't need to cook. Dinnertime would be the test of my cooking prowess and the ability to get several selections ready simultaneously. The main course was a pork steak, prepared on the waffle iron, and two or three other items which I proudly served. Eagerly I watched, anticipating the accolades that surely would come for this stupendous accomplishment, while Gary dug in with gusto and polished off most of his dinner before looking up at me. I was in tears. "What's wrong?" he said.

"You don't like it," I sobbed.

"What do you mean? I'm eating it!"

"But you didn't tell me you liked it." After traveling at least 10,000 miles, going home to mother was not an option, so somehow we managed to overcome this incredible gaffe on his part and my misunderstanding of his reaction.

Our rich dark wood floors were incredibly beautiful. Liberally applied YCO paste wax was buffed with half a coconut shell that must have been sawed in two to be so smooth. Loretta placed the cut edge on the floors' surfaces, put one foot on the curved top and, in a kind of jigging dance, rapidly rotated it as she hopped across the floors on the other bare foot, just like everyone's housemaid did. It was a vision of awe and a topic of conversation for all of us foreigners. She even did the red tile porch and steps that way. I was always afraid I'd skid on them when I wore heels, but they were so well polished they weren't slippery.

Almost immediately after breakfast, we heard loud noises that seemed to be a large amount of pounding on metal. It was the neighborhood's industry: across the street was a jeepney factory. Jeepneys serve as a sort of mini-bus even today and require stripping a jeep down to its chassis, then rebuilding it to accommodate several people. Each is gaily decorated with brightly colored paint. It has an awning roof for shelter from sun and rain, its edges frequently trimmed with lots of fringe. All the work at that time was custom and handmade, no mechanization that we could detect. The factory must have had a radio because we heard musical ads for zippers ("Talon zippers are the best for zipper quality.") and rum ("When people get together, they drink Manila Rum. At work or play together they drink Manila Rum. Manila Rum, Manila Rum, let's drink Manila Rum" in English.) Most other offerings presumably were in Tagalog,

Soon from somewhere behind our house came the unending strains of "Hi Lily, Hi Lily, Hi Lo" from the home of someone wealthy enough to own a record player. That one record was played over and over and over again. For a couple of hours late in the afternoon, we had relative peace and quiet. The factory shut down and "Hi

Lily" was turned off. But right after dinner we quickly learned what our evening's entertainment would be. One radio in the neighborhood (perhaps the same wealthy owner of the record player?) loudly shared what appeared to be extremely dramatic shows in the Tagalog language. We tried to read, but finally gave up and listened to the compulsory emotional offering. Looking up at our ceiling, we watched geckoes lap up a delectable feast of insects basking in the circle of light from our lamp. The other main appearance by geckoes was in the kitchen when they scooted across our dishes. Seeing a tail frisking rapidly across them was always an unnerving experience and usually elicited a tiny shriek from me. But they did help keep the insect population to an assumed minimum. The dishes definitely were washed both before and after meals and never got left in the sink. For the same reason, all opened food items (like flour, sugar, cereal) were kept in the refrigerator where nothing animate could get to them. The importance of this cautionary step was brought home to me on a weekend when Gary was gone, and a rat died in an under-counter cupboard which fortunately contained nothing.

Many Filipinos spoke Spanish, a holdover from earlier times. Tagalog was the principal native dialect or language used in our area. It certainly was unusual to see signs using two-letter words with no vowels, like "bk." Igarot was another main native tongue, but it was used principally in the uplands. We were told there were over one hundred others. I doubt I would have been able to differentiate.

Sometimes we'd see animals, usually chickens scratching in the dirt, in our street out front. But there were larger creatures, too. The first time I saw one it was somewhat far away and I assumed it was a dog because of its tail. But it wasn't; it was a pig! Filipino pigs had doglike tails rather than the curlicues seen in this country. And just what was that big pool of liquid at the edge of the street? And how did it get there? Speculations seemed the only answers, and none of them were pleasant.

Gary had taken a ten-day leave, so we quickly explored the air-conditioned Ship's Store and Commissary, then decided the swimming pool was the next best place to cool off. The water was about the same temperature and humidity as the air. The Officers' Club did have indoor air conditioning for the restaurant area. Watching movies shown at the Club was done outdoors. One memorable evening we saw "Quo Vadis" in a driving rainstorm that considerably distorted the action on the screen and nearly drowned out the sound. Since Dean and LaVon now had their recently arrived car, one day we drove with them to Lake Taal, a resort high enough in the mountains to be cool, and enjoyed the view of an island in a pond in an island in the lake, which of course was in an island in the ocean.

While everything we saw was exotic, unusual, beautiful, colorful — and occasionally disgusting — it was soon apparent that some of us were remarkable oddities to the local populace. The shipboard sunshine had bleached my already blond hair nearly white, and with my tanned skin they may have thought me a mutant. If I sat on our front porch, people would line up along the front fence to point at me and make remarks in inflections that seemed to express awe. Poor little Gregg North: a beautiful little boy with blond curly hair, fair complexion, and pink cheeks, he was just so fascinating that many Filipinos reached out to touch that hair. His tolerance level was very short so he'd turn to his father to be lifted up out of their reach, thereby arousing even more commentary — with Dean's height that little boy was so high he seemed to be almost regal.

During this ten days (which I soon learned was an unusually long time for Gary to be home), we also visited various families so I could become acquainted. Besides these casual visits were some social activities — a baby shower for Dottie Flood, who had just given birth to daughter Nancy, gave me a chance to meet many of the wives. Dottie's husband Frank was one of the other men who came to the Squadron on the same set of orders as Gary. Another event was a Squadron wives' luncheon at the O" Club. The only name specifically remembered among the many was Mary Jane Edixon. I don't know why, because there were quite a few of us there. Perhaps my mind was somewhat clouded: We sat outside around a low table maybe three or four feet in diameter nearly covered with a punch bowl. It contained a wonderfully cooling and delectable drink with lovely fresh pineapple pieces floating on top. Neither LaVon nor I drank alcoholic beverages — usually. I guess we should have been alerted as to this punch's potency when during a refill process a bottle of vodka was poured into the bowl. By that time, what was in it was immaterial. The cups of this liquid were so refreshing we had many, theoretically unfazed while we enjoyed becoming acquainted with the ladies in what seemed to be increasingly intensifying light and livelier and livelier conversation. Then we stood up — to experience a dizzying new sensation. I think we went inside to the air-conditioned restaurant for a lunch of some kind, and then somehow made it across the street to the swimming pool where we were to rendezvous with our husbands and Gregg.

Speaking of liquids, the greatest discovery of our entire stay was calamensi, a citrus drink similar to lemonade or limeade, but tastier and more thirst-quenching. A couple of times since then, we've been able to find the fruit here in the US. Each time, after three hours or more of squeezing the olive-sized fruits in a garlic press, I've been able to extract a scant cup of juice. I mixed it with a sugar-water syrup, and it was so good, evoking many, many memories.

Meeting so many new people was interesting, fun, and intimidating. I'd worked where I knew no one. My parents had moved far from my childhood home, so I never saw my high school friends. Other than my shipmates, I'd not been around anyone of my age group being married, not to mention having babies!

Military life was totally unknown to me, and I was concerned about making social flubs in its rather structured environment. Well aware of my own naivete', everyone seemed older and wiser, and I quite young and inexperienced. Having been away from the very active social life of college for more than a year, I was eager to be with people, even through those insecurities. As my current longest-term acquaintance in this new experience, LaVon was a sort of anchor, but for both of our sakes I did not want to become a chain. Fortunately most people were quite kind and welcoming. A Squadron couple, Jinny and Barney Martin, even invited me for dinner one evening when Gary was gone.

While many of the other Squadron people lived in compounds patrolled by guards and surrounded by high cement walls with broken glass imbedded in their tops, several did not, like us. Interestingly enough, most of the numerous stories about thievery seem to have occurred in the high-walled compounds - wallets stolen from under pillows or from shirt pockets hung on a chair beside the beds of sleeping people, even walls sawed open and refrigerators removed.

The wonderful ten days of leave came to a close all too soon, and immediately Gary left for who knew where or what or why or when he would return. (It turned out to be another ten days this first time.) I think perhaps Dean left, too, but not necessarily for the same length of time, and LaVon invited me to spend a night with her. Both of us may have been somewhat intimidated by the "alone" factor. I certainly was because of the open nature of our house and proximity of the neighbors, none of whom I knew nor could converse with. The thievery stories didn't set my mind at ease much, either. At any rate, I appreciated forestalling being alone for at least one more night.

Quite soon I squared my shoulders and decided being fatalistic was the only way I'd survive. Since I couldn't change the situation, what would be would be. Staying alone was the only option, and I needed to conquer my fears. Although I felt quite isolated — what English-speaking person I'd met lived closest to us, and where and how close was it? — Our only theft incident was one evening when several people came over for some socializing. Gary was gone. Earlier I'd washed out some lingerie and hung it on the clothesline off our back porch. Since the cotton took forever to dry, it was left on the line to be retrieved after guests were gone. When I went out, it had disappeared. Gulp! Struggling to quell momentary panic at the thought of someone coming over the fence, I reasoned that the colored panties with a day of the week on each (remember those?) were just too tempting for our next door neighbor — I hoped. There didn't seem to be anything else I could do but arrange my knife and cracker balls and go to bed.

Actually, it really didn't take long to accept staying alone at night. Theoretically for security, the open pocketknife was kept on a small table beside the bed, right next to the cracker balls. The knife probably would have been more hindrance than help should I have had an invasive visitor. Fear surely would have paralyzed me. But if I could move, perhaps the B-B type cracker balls, which made a very loud noise when thrown against a hard surface, might frighten away an intruder. When Gary was home, the noisy balls supposedly would allow us to get up in the dark to get to the light switch without stepping on the huge water bugs that we called cockroaches. But first, before retiring, we liberally sprayed the floor with an insecticide that had an extremely sickening smell. It must have been the only one available or we wouldn't have continued using it. In the 1980's there was a popular perfume some women chose that I thought smelled exactly like that bug spray! When those wearing it passed by, their scent evoked unpleasant memories of those ugly bugs, certainly not leaving the effect the women imagined.

Wives participated in the social activities without husbands, even if they happened to be the ones hosting an event. I was excited about an upcoming party at the Squadron Commanding Officer's home and busily made a new dress for the event. Gary was gone, so Dean and LaVon picked me up to go there. Probably the Commanding Officer, Lieutenant Commander Hall, was gone, too, as Gary was frequently in his crew at that time. There were lots of us there that night. The party progressed, and eventually Dean and LaVon wanted to leave. Apparently Pauly Varner was aware of my apprehension about staying alone and invited me to spend the night with her since husband Stretch was also away. She lived on the Base, had an extra pair of pajamas, and probably an extra bedroom. Somewhat hesitantly I said okay because I didn't really know her. The North's left, feeling I was in safe hands. As the party was breaking up, someone said, "Hey, let's go to State College!" Pauly was among those who agreed. My apprehension mounted even higher because now I was in a totally unknown situation, with people I didn't know, having to go to a mysterious destination, and timid about asking someone to take me home. "State College" was actually Sheran Cove, a bar located just outside a checkpoint at the edge of Cavite. The checkpoint was designed to keep out invasions by a terrorist group, the Communist Huks. This evening was becoming more and more thrilling. But off we went, and I had still another Coca Cola. We were there awhile and then went to Pauly's. She'd also invited Joe Schmidt and Fred Groenert, whose wives had not yet come to the Squadron. We had a nice conversation, they left, and the evening ended.

Next day LaVon called to hear about what happened, and after my relating the activities said, "Did the guys spend the night?" Internally I was very insulted. While I was naïve, I wasn't stupid, and I knew rumors could start over comments like that. Fortunately, as far as I know, LaVon was the only person to voice it.

I can't seem to remember shopping for food and transporting it back home. Actually, I didn't really know how to shop for a twosome, especially when one was gone much of the time, but I didn't figure the learning curve could be too high. When Gary was home, there was no problem. We rented a Lambretta scooter with a basket on the front that held a couple of bags of groceries when Gary was home. We had bicycles but, never having had one as a child, I didn't feel confident enough to go exploring by myself. So I spent a lot of time alone. I've never had any problem entertaining myself. I'm sure I read a lot, did some embroidery, probably wrote lots of letters to family and friends, and tried to ignore the irritation of the interminable "Hi Lily" refrain.

When Gary was home, we usually stayed in the area. But once we had dinner at the Port Officers' Club in Manila. Another day we were in Quezon City, near Manila, where I bought silver filigree earrings made by Filipino nuns. For our mothers, Gary's sister, and me, we also purchased pina cloth handkerchiefs. Pina cloth is very delicate and woven from pineapple leaf fiber. If restructuring a jeepney by hand was painstaking labor, it's hard to categorize the delicacy and difficulty of making these two objects of Filipino art. The metalworking skill to make thin silver wire, then bend and twist it into intricate patterns, boggles the mind. So does construction of the handkerchiefs. That would have involved extraction of fibers from the pineapple leaves and conversion to thread, then weaving it, and finally hemming and decorating with fine embroidery. I suspect the shopping and dinner may have been the same day as the trip to Lake Taal, but perhaps not because I seem to recall a boat ride to the dinner. Other than that and our scooter tour of Cavite, we did no sightseeing. The leave, followed by Gary's ten-day mission, took a big chunk out of the two months' stay in those tropical islands. Besides that, we were quite careful about our money, looking ahead to Gary's return to school.

Military hierarchical structure was interesting to observe. Date of rank was the requirement unique at Sangley Point to be granted Base housing. We never would have qualified. And, although it was not an actual barrier, there did seem to be a slim silken curtain silently adhered to by some of the men who were career (Regular) Navy and those who were not (Reserve). It was less obvious among the wives. And among the men were a few who seemed a little more conscious of their rank. If an Ensign (Gary) was "the lowest form of human life," then it seemed his wife had to be beneath that. While most people were pleasant, there were some with whom I don't recall ever exchanging a word.

Surprisingly I did know one person in the Squadron. Before I came to the Philippines, Gary lived in the Bachelor Officers' Quarters (BOQ) along with several other unattached males. Their common cocktail hour activity was watching sunsets over Bataan. One evening Gary and another fellow had the "Where are you from?" type conversation. The other guy was from Spencer, Iowa, my hometown! The high school was small enough so everyone knew who everyone else was, even if not in the same class. It was Glen Carlstrom, a couple years ahead of me, who was active in the school's athletic programs and school band, so of course I knew who he was. In the Philippines, Glen had met a girl whose stepfather was a government official living in Manila. Glen and Marilyn would marry soon after my arrival, and Gary became one of the holders of swords (Honor Guard?) crossed over the newly-married couple, just like those romantic scenes we saw in movies and newsreels. The anticipation of this splendid affair had everyone excited. Naturally I would need a new dress and designed one I didn't think I could make in time. Then I was able to locate a seamstress who would tackle the job if I would bring my drawing and the fabric. This necessitated a trip to the open air market.

The weather was always hot and humid in almost exactly the same proportion. One needed a very small learning curve to know that in those pre-air conditioning days. I had been warned the market was definitely not a place for pregnant visitors but, since I wasn't, the prospect of the visit had seemed exciting. (Anticipation is nearly always better than actuality.)

Gary and Nancy at the Carlstrom's wedding

It hadn't taken long to learn the term "sanitary conditions" was little practiced in this different land. With some reluctance on his part, Gary agreed to our adventure. He'd already been to the market. Recollection is unclear as to which attacked my senses first — sound or smell — but the aroma seemed even more pervasive than the rapidly swelling cacophony. The market was roofed with some kind of cloth that filtered the sunlight, but it surely could not ever be called a "fresh air" market, and it was immediately noticeable that "smell" translated into stench. I was reminded of a much-used wooden outhouse on a bright July day on my uncle's Iowa farm, only more so because this market obviously never was scrubbed. The sight sensation was delayed. Tropical sun is intense and it seemed adjustment to the interior took longer than customary.

We strolled more slowly than I quickly discovered I really wanted to, but I was looking for just the right piece of fabric. Gary was looking for a beautifully woven and embroidered shirt called a Barong Tagalog, traditionally national formal dress attire. While some were made of pina cloth, they were incredibly expensive, and even those of inferior fabric possibly available in the market were not cheap, but highly prized. The shirts were extra long and worn loose outside the pants. The neckline plackets were left unbuttoned, and much of the surfaces in the upper part of the shirts were intricately embroidered.

The aisles were filled with wooden tables of varying sizes, heights, and stability, holding a wide variety of — dare one say it? — Goods. Colors of all vivid shades were displayed everywhere like a patchwork crazy quilt, some against the muted background of skin in various hues of light brown. Countenances were smiling, frowning, and mouths open wide shouting wares or just jiggling up and down in conversation with neighbors or customers. The vendors called their wares in Tagalog, which here seemed to differ only slightly from the clucks of the overly prevalent chickens, both those caged and others free to roam at will. Did we walk on packed dirt or was it cement that was emanating more heat? Whichever kind was bordered on each side with small alleys, theoretically to carry off varying animal and human effluvia. Darting in and out everywhere were the omnipresent little boys whom we called pea-pickers, dressed only in T-shirts. A much feared disease in the Philippines was tuberculosis. The T-shirts, as well as hastily improvised cover for people's heads in sudden rain showers, were thought to help forestall onset of the disease.

A truck backed in towards us with a roar that accentuated the heat which his motor added to the mix. Leaving it running, a muscular young man jumped out, went to the rear of the truck, and onto his brown back, beaded and running with sweat, hoisted a side of what presumably had been bovine a very short time before, judging by the fluids we noticed. A hovering cloud of flies - or whatever they were - which had been disturbed in their consumption of the meat, quickly descended and blanketed the next piece. The man found the right stall and delivered his load while eager purchasers decided which choice cut they wanted. This was repeated until the truck was empty and buyers' baskets were full. It was one of those fascinating visions as horrifying as it was hypnotic. But as the next truck pulled up, we tore our eyes away and managed a rapid detour and retreat.

Eventually we found the desired sky blue fabric, quickly had the proper length cut, and escaped, the Barong Tagalog purchase to wait for another day and location. As we departed, the rich stew composed of the sensory elements insistently followed us as long as memory. (The desired Barong Tagalog was finally acquired as a gift from our son-in-law in 2005, purchased during one of his business trips to the Philippines.)

Another memorable event was attending a cockfight. Bob Hansen and Don Burdg went with us to a sort of barnlike structure filled with wooden bleachers arranged around the fight pit. The betting was most interesting. A broker walked back and forth on the ground floor. People shouted down at him or at the pair of fighting cocks, often in a rhythmic "a bese, a bese, a bese." At the end of each fight, the broker would start a money path hand-over-hand through the crowd to the winning bettor. He must always have had the correct memory as to whom and how much because we never saw an argument. The birds were fitted with razor-sharp curved knives at least three or four inches long attached to their legs. One wonders how they got their training because nearly all the fights were to death. Usually the winners were very aggressive, but one exciting bout extended to nearly unbearable length. It was especially captivating because the one most severely wounded was unwilling to lie down and die. He valiantly continued to attack the apparent, but now hesitant, winner. That one almost seemed to have been sickened by the damage he had done to his opponent and wanted nothing more than to escape the whole vista. Every object within sound vibrated with the verbal support from every throat for the bravery displayed by the underdog. His refusal to die despite grievous injury, his instinct to become the aggressor, was brutal and elemental, magnificently thrilling — and repulsive. While I'm glad I experienced seeing this cockfight, it is highly unlikely I would ever, ever have gone to another one.

When we left, we saw many pots boiling over fires, presumably containing the losers. Differing chicken delicacies obviously were being prepared and offered for the discerning palate. For some reason that seemed more sickening than what we'd just witnessed. Don Burdg bought a balut, a Filipino delicacy. When we got to Hansen's' house he opened the egg on an outdoor table and sure enough, there was a baby chicken, feathers and all. We understood it all was supposed to be eaten raw, including the feathers. The day had been quite a test of what the stomach could bear!

A different kind of unpleasant event happened one evening when Gary was gone. Someone offered to drive me to the O' Club for Bingo. I thought it would be a nice diversion. Awhile after arriving, Cathy Hall, the CO's wife, came up to me and in a taunting, singsong manner said, "I know where your husband is tonight." I snapped at her, "Well, don't tell me, I'm not supposed to know, and I don't want to endanger his life!" Talk about an oddity! Normally shy and not very assertive, my snapping at her shocked even me. I thought, "What have I done?" Although my private evaluation of her was not particularly high, I owed her the respect to which she was entitled by her husband's position. And then, one must consider the all-important Fitness Report which haunted each man. She looked at me in surprise and walked away. I was so angry! She wasn't supposed to know any crew's destination, including that of her own husband, and the fact that she said she did was frightening. Although I didn't know what was going on, we all knew it had to be dangerous, and talking about it potentially could heighten the peril. I've held that resentment towards her for years and finally decided that perhaps she was testing me, and perhaps even had been told to do so. While I doubt it, I'm finally giving her the benefit of the doubt.

Not only did Gary not talk about what his primary job involved in the Squadron (pilot), he didn't talk about his secondary one (in the Supply Department and as Accounting Officer in the P.I. At the end of his tour with VQ-1, he was the Supply Officer). It was many years later I learned that while in Japan serving as Navigation Officer, he went on the first operational mission of each new pilot who rotated into the Squadron. That, coupled with the frequency of flying that Commander Hall liked to do (including volunteering for unusual missions), and filling in for other pilots where needed (as each pilot did), he seemed to be away from home more than other pilots, or at least it seemed so to me.

Perhaps one other time, undoubtedly I was considered to have said something out of line, although I certainly didn't think so. I interviewed for a teaching job in the dependent school on the Base. During the interview, the officer in charge asked me where I'd gone to school.

"Cornell College in Mount Vernon, Iowa," I responded

Actually kind of smirking he said, "Well, that's a lot of borrowed names, isn't it? I went to the *other* Cornell — Cornell *University* — in New York."

Very matter-of-factly and truthfully I responded, "I think if you'll check, you'll find Cornell College preceded Cornell University." (It did.)

His face darkened, the smile disappeared, he concluded the interview, and I didn't get the job nor even a courtesy call to tell me no.

When I learned we were to live in the Philippines, I'd bought some fabrics to bring along overseas for new clothes, speculating they'd be quite unique. One became a skirt, another a dress perhaps nice enough to wear to parties. On one of our first trips to the Base, a fat little Filipino man rode past on his bicycle wearing a shirt made from my skirt fabric! And every time we went thereafter we saw him again in the same shirt! As for the dress, that was another experience. Not only had I become accustomed to living and sleeping alone, one day I felt brave enough to take a jeepney alone to a no-longer remembered destination. It was a Monday, unfortunately the day the prostitutes visited a doctor to see if anything unwelcome had been deposited over the weekend. At one of the first stops, one of the working girls got in my jeepney wearing a dress from the same fabric as mine! Wo-o-o-oo — so much for exclusivity! Fortunately I was wearing something else.

The men had tropical hours. This meant that if they weren't flying, starting work on the Base began very early and working until 12:30 gave families all afternoon and evening to be together. Those infrequent days seemed like we were on vacation. Usually we went to the pool, but one day we hopped on our rented Lambretta and took off to sight-see. A famous (or infamous) place we observed was called Officers' Beach snidely, but of course unofficially. To call it disgusting would have been a compliment. It was a tidal pool partially edged by the back-ends of rickety houses on stilts. Always present in the water, which varied in depth according to the tides, were two men searching for crustaceans to eat. The aroma from the area was quite intense, leading to the most probable conclusion as to the plumbing system in those stilted houses, especially given our personal experience with the neighbors' houses.

The only time I knew where I was in Cavite was when we were on a road next to the water, going either to the Base or in the opposite direction to the North's'. It was fun flying along on the back of the scooter, wind in my hair, passing Filipino houses open to the sun from which children ran to the edge of the street shouting, "Hi, Joe!" Also what might be seen was a person sitting in a chair with another person searching the hair for various insects, not nearly so pleasant a thought to contemplate. Still another interesting sight was an occasional bene (sp? pronounced like Benny) boy mincing along on his high heels, sporting an attractive dress and bright red lipstick. It was my first known sight of a homosexual. Supposedly it was considered a Filipino honor for one to live in your house. A fascinating speculation at an All-Hands party was the date of a very young sailor new to the Squadron. Did he really know what he had, if that's what he/she was? We worldly twenty-two and twenty-three year olds wondered?

Most days were sunny with the 90+ temperature and humidity about the same. But usually around 4:00, there would be a sudden downpour lasting just a few minutes. One day there was a most unusual event: lightning and thunder. The first clap was so sudden and unexpected that Loretta, who was ironing, screamed and leaped across the room, iron in hand. My first thought was that the house had been struck by lightning that had traveled through the cord to the iron, throwing her across the room. But she was just startled and frightened.

We'd been told that if clouds were on the other side of the silently lurking Bataan Peninsula, there'd be no rain. But if the clouds moved to hide Bataan, we would have a shower. So when a day dawned gloomy, with Bataan sulking as though miffed about its reputation for that World War II Death March, and with even more oppressive sultriness than usual, we knew something different was happening. The wind got stronger and stronger and rain became nearly constant. Luckily Gary was home because we needed more than one person to hold buckets for the leaks entering the house in different places. The most memorable one happened in the dressing room: the wind was so strong it forced the rain through a minute hole under the window in the outside wall into the very center of the room. In just moments we had to exchange receptacles, with scarcely enough time to empty one before we needed another, time after time. This, my first typhoon, was frightening! When the worst of the storm was over, we went outside to survey the damage. We'd lost five banana trees and some pineapple plants. None of the fruits were ripe yet, so it was disappointing no longer to be able to anticipate sampling several different varieties of bananas. We hadn't realized there were so many kinds.

I felt so bad for Gary — I could not stop my stupid crying whenever he left. He might go to work at the Squadron, then come home early to sleep and pack. That meant he was leaving. I feared I'd never see him again, fed by my apprehension about flying in general and the unknown dangers I felt he must have. So the flood of tears would begin, despite all my desperate efforts to stop. I don't recall that it was for me so much. I knew I'd be alone the rest of my life, but thought that could be handled because I was already pretty much alone a lot of the time. It was that he had so much to offer others and wanted to help them. He'd planned to be a physician after a childhood medical problem put him in the hospital for seven weeks. He wanted to be a small town physician who would make calls at any time, any place. When near the end of his seventy-nine Squadron missions, he decided he'd already been away from home more than he wanted to, so he switched his goal to dentistry, still in the health field. That fear of mine about his flying expanded to include anyone in our family who was flying, and it lasted many, many years.

One day, not long ago, I sat myself down and said, "This has got to stop! You can't enjoy going anywhere nor having anyone come to see you. Enough is enough! You've taught your children that, `If a situation is intolerable and can't be changed, you have to change your attitude.' So now it's time to change yours!" Incredibly it worked and now I actually enjoy flying! That is, if there aren't too many bumps and sudden altitude changes.

Very quickly September arrived and left. I'd been in the Philippines less than two months when word came that the Squadron would be moving to Japan. Wow! It was October. Eternal summer would be gone and we'd need winter clothes! And blankets! Quickly I sent off a letter to my folks to get them shipped to our new address. In the meantime, at our request, Gary's parents had bought us a stove, refrigerator, and wringer washing machine that had not yet arrived in the Philippines. They came just a few days later and the decision was made to keep the washing machine, not usually included in the base housing we would now qualify for in Japan. Gary had left for NAS Iwakuni to help make arrangements for the Squadron's move there, and I would not see him again in the Philippines. So I put out the word there was a refrigerator and stove for sale. The refrigerator, made by General Electric or General Motors, was sold immediately. At that time Filipinos so loved General McArthur that anything with "General" in its name or make-up was instantly desirable. The stove, however, was an off-brand. Our landlords, Mr. and Mrs. Abarro, wanted to take a look at it for the house and so came there one evening. They'd not been in the house since we'd lived there, although they'd been outside to harvest some of the useable produce that survived the typhoon. They brought us a pie made from those hanging gourd types in our front yard which were called patola. As I recall, they were a white sweet potato and the pie tasted something like coconut.

They sat to visit a while and then decided not to purchase the stove. It had an inflated price, much more than its worth, which they recognized. I'd felt kind of sheepish asking that much, but it was what I'd been told to do by people who'd been there longer. Before we left the Philippines, however, it was sold. With the sale of it and the refrigerator, the two appliances exceeded what we'd spent on all three. Somehow the conversation with the Abarros turned to World War II and the atrocities committed by the Japanese. The Abarros had seen the actuality of the terribly sickening stories some people thought were myths about throwing babies into the air to be caught on bayonets. Mrs. Abarro was probably not much older than I, if not that old. Very matter-of-factly she told how she had been hidden, but in a place where she could see what was happening, when the Japanese came to her home. She watched as her mother and older sister were raped and killed. What can one possibly say after hearing those firsthand experiences? Japanese visitors were still not welcome in the Philippines.

Those of us who lived off-Base made arrangements and supervised packing up our belongings, giving up our homes in preparation for transportation to Japan.

Surprise! We couldn't go! Japan was having a huge typhoon! Now what? Where do we stay? Fortunately there were several families still able to remain in their Base housing until the last minute, so we were lodged with them for about a week. LaVon, Gregg, and I stayed with Pauly. I think I may have slept on the couch in the living room because that's the only room I remember there. LaVon and Gregg had the use of a bedroom. Neighbors to Pauly were the Ebersbaker's. Their time left in the Squadron was so short they would be staying at Sangley Point and involved in whatever the Squadron needed done after its official departure. In preparation for leaving, their extensive liquor supply needed clearing out. So each evening we strolled across the lawn and were treated to Bacardi's, which I've never had since, although they tasted very good. Other than that, we probably just stewed and fretted about when we'd get to leave.

At last we climbed aboard a MATS (Military Air Transport Service) plane with canvas bucket seats, some running parallel to the outside walls, a few in the middle facing forward. The plane was filled with many children, wives, and a few enlisted men. Hurray! We were on our way to Japan for another new experience!

Land of the Rising Sun

Nijimura = Rainbow Village

I'd been on Guam, Luzon, and now would be landing at another island become familiar by way of World War II - Okinawa. The approach to Okinawa was not pleasant because the plane circled the field for a seeming eternity. We had been traveling for more than 900 miles. At last we landed and were able to stretch our legs during the forty-five minutes we were there. Starving because our box lunches had been prepared for noon and were waiting for us on Okinawa where it was now about 4:00 PM, we gulped down some food. This alleviated our attitudes for

The family parade

awhile. Then off we went again, heading for the largest Japanese island, Honshu, to be our next home.

Now we had another 600-plus miles to travel in this long, long day. Excitement mounted as we approached our destination. Tired and hungry kids were crying, some were getting sick, and all of us were impatient and cross. When our husbands learned that we were circling the area, they managed to get it stopped. So at long last we landed to meet our husbands who'd been there for at least a week. The first order of business was food, a decent hot meal being top priority. It was nearly 10:00 by the time we got our food. Then we split up, some families to remain in Iwakuni, the rest of us to take a boat for the two-hour trip across the Inland Sea to the Army post at Nijimura, beyond Kure, where temporary housing was available for us newer Squadron members. Joyfully anticipating a horizontal position on a mattress, we were met with the last straw of this extended day. Though it was now into October, there was no heat, something we'd definitely become accustomed to in the steamy Philippines; no efficiently categorized Bedding, blankets; Bedding, sheets; Bedding, pillows; nor Linens, towels, all of which had been arranged to be supplied upon our arrival. While we dependents waited shivering in our new homes, several angry men marched to wherever such incidentals were kept and returned with armloads of said supplies. Sleep was merciful.

The next morning we did find interim pots, pans, dishes, and flatware, but I don't remember how, what, or where we ate. Nor do I remember how soon we received our household goods from the Philippines. Our supplied stove was a joke: if an electric burner was turned on immediately upon arising, by the time bathroom activities were done, bed made, clothes donned, etc., it was possible to detect a vague warmth by placing one's hand flat on the burner. Thus began another of the battles with the Supply Sergeant, quickly named Little Caesar. He possibly was resentful that this whole bunch of people - who weren't even Army! — had been dumped on him and kept demanding all sorts of unpleasant and unreasonable things, like a stove that worked.

This Army Post was small, mostly residential, and located between the mouths of two rivers emptying into the Inland Sea. Gary and I felt we had the absolute prize location for our quarters. The buildings in our section were four-unit ones, the center two units side by side, and each end one facing outward with backs to those center units. We were fortunate to have an end one facing out to the Inland Sea.

Contrasting strongly with the riot of color of our recent residence, our living room seemed drab despite a deeply wine-colored rug. What wasn't wood on the serviceable mission-style furniture was covered in a dead gray. Perhaps that first night set the feeling, but the combination living-dining room always seemed cold to me, and I could never make it feel welcoming and comfortable. Despite numerous windows, strangely the first floor never seemed light. It was winter while we lived there. Days were shorter, and there were many more cloudy days than sunny ones. After a few months I purchased some wildly abstract fabric to make curtains. I didn't particularly like the fabric, but it did provide some color. Without any regret whatsoever, the curtains were left behind when we moved to Iwakuni. Despite its perceived shortcomings, though, we were grateful to be together and in government-provided housing rather than having to find a paper Japanese house and shivering beside a hibachi for our only heat and cooking.

But oh my! What an amazing view we had from our second-story bedroom! We had windows to the south, the east, and the west, this latter through an open closet door. This room was definitely our favorite. We bought a couple of basket chairs for the corner, and I could set up my sewing machine on the vanity that sat between windows. The rising sun (appropriate!) came up over a small terraced mountain visible from the bed. The south windows offered a lovely Japanese "painting," the sea bordered with outcroppings of various types and heights of land and occupied by occasional moving objects like fishing and other types of boats. I don't know why we never took a photo of it, it was so beautiful.

It hadn't taken long to become acclimated to the heat of the Philippines. But the temperature here was a distinct shock, especially in the apartment. If outside was sixty degrees or below we had heat, if above we didn't. This meant that we daytime residents usually froze all day in house temperatures close to that magical sixty-degree mark. But by the time the husbands got home at night, the house was so hot they complained (and couldn't believe our complaints about cold!) as they threw open doors and windows.

Again we had a maid, and she was a very sweet girl, occupying her time when work was done by crocheting doilies as gifts for us. Our verbal communication was almost completely nil. I really felt badly about letting her go, quite cognizant that she was probably supporting her family in that predominantly manless society following the war. However, it was really boring when Gary was gone, again often. Between that and his spending four to five hours a day on a boat trip to and from the Navy Base at Iwakuni on a regular work day, which definitely lengthened his away time, there just wasn't much for me to do. I felt uncomfortable having to move from room to room to stay out of her way while she cleaned and somewhat guilty, too, to be sitting there reading a book while she had to work. Besides, as a

Japanese Traffic ballet

new bride, I still wanted to make a home for my husband, even though I was reasonably sure there would come a day in my life when I'd kick myself. So when our washing machine arrived from its journey from the Philippines, I bade our maid a sad farewell. We'd finally triumphed over Little Caesar and had a stove that worked. Now I was ready to be a true wife for my husband.

I was not terribly experienced with a wringer washing machine. The first day after ours came, in the kitchen I set up what I assumed was a convenient layout. Things went well until a hand towel got wrapped around the wringer. I tried to grab it, but the wringer grabbed my fingers instead. I kept trying to pull my hand out as the wringer inexorably moved up it, my wrist, my elbow, nearly to my shoulder! I couldn't reach the handle to turn off the action since it was on the opposite side of the washer. I hadn't realized there was a bar across the wringer top, designed for just this sort of situation.

When I found it, I was able to pull out my arm and examine it carefully. It didn't look too bad, but my hand was a totally different story. It seemed to be swelling. The skin and blood vessels had all been pushed toward my fingers. I grew uneasy. By this time I'd become pretty good friends with Marilyn Carlstrom, who lived at the other end of our building. I walked over there and told her about the morning's activity. She was shocked at how my hand looked and said I should see a doctor right away; she'd go with me. Most of the working parts of this Post were on an island near Kure, our nearest large city. The island was Eta Jima, the former Japanese Naval Academy. We walked to the main street outside the Post and caught a train, then a boat to Eta Jima, and eventually found where the medical facility was on the island. I don't remember too much about that trip but think the doctor told me the swelling would probably go down in a few days. So back home we went.

Several days later Gary came back. When he landed in Iwakuni, someone asked, "Did you hear what happened to your wife?" and his heart started thudding. Immediately he phoned to find out. When he got home he was shocked, too. The swelling had not gone down. It now had expanded to become a big black solid puff about an inch high covering the entire top of my hand. It made my whole arm ache to the bone. We decided it definitely needed some attention, so we repeated the journey Marilyn and I had made. The doctor told us it was a blood clot and he'd drain it as much as possible. He lanced it and started pressing out the dark, dark, nearly solid "jelly" which my blood seemed to have become. After a while he happened to glance up at Gary, who had turned green and was just about to collapse. Then he got the attention. When the doctor returned to me and got out as much more as he thought possible, he forced a long strip of gauze into the incision to soak up residue. Far more painful than the previous treatment, it stayed in a couple

of days, and then we pulled it out when I could no longer stand the throbbing. That was "Nancy's Big Adventure" with being a housewife.

People who do not read for pleasure miss so many interesting experiences. It was a very rewarding time-user for me. Case in point: Gary was gone; I knew not when he'd return, and I was reading a well-known best seller. I have never cared for scary movies, novels, etc. but was somewhat intrigued and perhaps a little tired of my usual fare. The book was ***Something of Value***, about the early 1950's Mau Mau uprising in Africa led by Jomo Kenyatta. Hypnotized by the violence in the plot that was developing, I read late into the night and finally fell asleep. Suddenly I felt a hand on my shoulder! Oh, my! The Mau Maus were here! I was under attack from a home invader and would undoubtedly be killed! My "attacker," of course, was Gary, and I had been so sound asleep I hadn't heard him say my name. He said my eyes were really as large as the proverbial saucers when I turned toward him. Now, that is realistic writing! This 2:00 AM appearance was an unusual time for him to arrive. Someone had a car in Iwakuni and Gary had ridden home with him.

Even though I was somewhat bored, it's not because there wasn't socializing. I really appreciated being close enough to people to get acquainted, and the isolated feeling disappeared most of the time. No one from the Squadron was in the apartment next to us, but between it and Carlstrom's lived Stretch and Pauly Varner. Again Pauly came to the rescue when Gary was gone, sometimes with a meal (she taught me how wonderful duck is), and it was fun to step out on the back stoop and look over to someone I knew looking back. I really didn't know Marilyn in the Philippines except at their wedding, but now found we had many things in common, among them being the two newest brides in the Squadron, and of course, Glen, slight as our acquaintance was. The Martin's lived directly across from Carlstrom's at the end of the next building. One time Jinny had a morning coffee for us wives. Each time someone would come to the door, her Kerry Blue dog Spook leaped up and raked it from the very top to the bottom. It was quite an amazing sight!

Next to Martin's lived Dean, LaVon, and Gregg. I spent considerable time there because we were now "old" friends. They kindly offered to share many a spaghetti supper. One of my favorite Gregg stories was one day he was on his back on the floor, wildly waving his arms and legs. When asked what he was doing, he responded, "I'm a cockroach dying!" The very best Gregg story, though, happened one Sunday afternoon when Dean was gone. The doorbell rang, I went to answer it, and looked w-a-a-y down to see Gregg. "Can Gary come out to play?" he implored. He could and did.

Beyond the North's lived Bob, Jean, and Patty Hansen. I think Gary had spent quite a bit of time with them before I got out to the Philippines. One memory is of little Patty's pretty brown eyes and blond hair. When we traveled across the Inland Sea that first night, she was sitting on Jean's lap across from us, and Gary got her finally to close those marvelous eyes.

Our backyard was enormous, shared in common not only with our two buildings but homes and apartment buildings along the streets perpendicular to ours, all the way to those at the far end of the block, surely farther than the length of a football field. Some of those were one-story, some two, some with several bedrooms for families with more children. Quite a few VQ-1 families lived there, as well as some Iwakuni Base employees not in VQ-1.

This Post seemed to be in a developmental stage with security appearing rather relaxed. Perhaps at least once a week, there were two Japanese people who searched our garbage, presumably for tasty tidbits. I don't know if they were maintenance-type workers or area residents allowed onto the grounds for this special treat.

One special memory about the back stoop was when we bought silk paintings by an itinerant Japanese artist. Four of the larger ones were of the four seasons; two little ones were of birds on bamboo branches; two other large ones were of a male and a female dancer, each in colorful garb. The artist, who was allowed to come on the Post, worked on the stoop framing our choices: black lacquer for the seasons, dark bamboo for the others.

Families left, some to NAS Iwakuni as housing opened up, some to new assignments, and some home to "The Land of the Big PX." New families arrived, but it's hard to remember who and when. I think Varner's, Carlstrom's and Martin's all left around Thanksgiving. Jim and Holly Carberry took over Martins' apartment, and Jim and Betty Rex moved into the Varner's

LaVon and I became adventurous. One day we left Gregg with their maid and ambled outside the gate to catch a train into Kure. I don't remember our errand, probably to shop at the department store. Our stop must have been near the end of the line because there was no one else already in the train or joining us there. The seats in the cars ran parallel to the sides. Since we weren't sure how many stops there were before arriving in downtown Kure, we sat down near the middle. Soon the train stopped and two people got on. One sat very close to me and the other very close on the other side of LaVon. She and I looked at each other uneasily. This became a pattern with every stop, the seating moving ever out from us while maintaining our near neighbors who seemed to be edging ever closer to us. We were very ill at ease and very aware we were the only Round-Eyes in our car. No one had told us how cozy the seating got on Japanese trains! At other times we accompanied Navy and Army wives, the latter to guide us around Kure for shopping, dining, and a good seamstress.

In the evenings, several of us got in the habit of walking over to the boat landing to meet husbands when we knew — or at least hoped — they'd be there. It became an opportunity to socialize while we waited for them. We often met a Japanese guard carrying a rifle. He always greeted us pleasantly with "Konbonwa," a word I later learned was "Good evening," but I was quite apprehensive at first. I remembered what that Rising Sun had meant and didn't like the looks of that rifle at all.

Not everyone did what Gary and I did, which was hurry back home, gobble down the meal I had ready, and run to the movie theater. If we were lucky, we didn't miss too much of the feature or maybe not any if the Added Attractions ran a little long. I quickly learned that dinner dishes could wait until morning even when we didn't go to the movies. If I finished them right after dinner, Gary would be asleep on the couch.

A big purchase for us was a record player and a tape recorder. On many of the weekends Gary was home, we would lug them catawampus across the open backyard and another block or so to the library. The library had been formed from the middle two units of a residential building like ours. Connecting doorways were opened to allow movement between them. One of the upstairs rooms contained many music albums which we taped for playing at home. There was quite a good-sized collection of phonograph records, but they could not be checked out. An additional huge advantage of the tape recorder became that we could exchange tapes with our families back home. It was wonderful to hear their dear voices and descriptions of familiar days.

We were fortunate that we almost never got really homesick, probably because there was so much new happening in our lives. The first Christmas was different, though. We became very nostalgic, once we got over the laughter on Christmas morning. We'd gone into Kure and bought our very first Christmas ornaments. While walking along a street in the business district, we heard a drum and some reedy instruments, then spied the source of the sounds. It was three men, one unusually tall, walking along single file as they played some barely recognizable Christmas carol. It was our very own Christmas parade!

After we got back to Nijimura and deposited our packages, we went to nearby Hiro to buy a tree. It was rather spindly, but we figured the ornaments were so lightweight ("Made in Japan") it wouldn't make any difference. We decorated on Christmas Eve, traditionally for me opened some presents then, and went to bed. When morning came, we were to open the rest of them, which was Gary's family tradition. We got partway down the stairs to where we could see the tree and burst out laughing. The boughs were bent over so far it looked like we should have turned it upside down.

We'd decided that since we had no family, later in the day we'd deliver a Christmas card and a verbal Merry Christmas to each family we knew as part of our own little celebration. It wasn't a good decision. With each door to a happy home, we became more and more glum. We didn't make that mistake again the next year.

I can't speak for Gary, but the only other time I became a little homesick was in July when visualizing miles and miles of the Midwest's beautiful fields of ripening golden grain and next to them the dark green of as yet untasseled corn.

Generally speaking, shopping in Hiro was an enlightening experience. The shops were all open on the street side. To keep down street dust, the operator frequently sprinkled the sidewalk or street in front of his/her shop. Since it was winter, there seemed to be no way we could keep our feet warm. We didn't want to shop late in the day because over the doorways, food shops hung squid with tentacles invitingly(?) hanging down. Even though it was winter, it was warm enough for the squid and various other foodstuffs to become very odoriferous by 4:00 p.m., perhaps not so much so as in the Philippines, but definitely noticeable.

The facilities at this Army Post were quite scattered. In subsequent weeks, several were moved from their locations to be centralized at our physical site. But on one of the first weekends after we arrived, we learned that the Officers' Club in Hiro was having a sale on brass. We decided to go to it late in the afternoon and then have dinner there. I really don't remember how we got to it, but it was certainly closer than Eta Jima Island. We must have taken a tiny Toyota cab. Anyway, we did get there and bought a lamp base and a couple of ash trays. We also met an Army captain and his wife who invited us to have dinner with them, then asked us to go with them to the British Command Officers' Club. (Britcom had a presence quite close to us.) We arrived and found a rousing party going on. The captain knew some of the Brits who said that even though it was now about 9:00 p.m., they were having such a good time they extended their Happy Hour. We discovered that their otherwise rather dignified demeanor was apparently just a cover-up for quite a jolly group! We sat with a man who ordered Scotch. When it arrived and he took a sip, he looked startled, then reached in the glass while roaring, "Oh, Gad, not ice!" and threw the ice on the floor. A little later, he and a lady friend did an amazing dance to "Cherry Pink and Apple Blossom White," covering the entire length of a quite long ballroom during an elongated note on the word "Pink." At the end of the length, he rose on his toes, bent over, and did a little bow that looked like he was about to vomit. It was quite fascinating, and we tried to hide our hilarity.

A very short British officer with several gold crowns on his shoulders came up to me, saying, "My Dear, you have the bluest eyes!" (Smooth talk, eh?) After I gave Gary a desperate look to rescue me, we had some further conversation. Then he said, "You must excuse us (he had another officer with him with nearly as many gold crowns). We have a couple of young ladies waiting for us at a hotel." I tried not to show my innocent astonishment that he should be so open. We did learn that this was probably a rather common occurrence because most of their foreign tours were four or five years long, and few wives accompanied their husbands.

We took a shopping outing to Eta Jima with several other VQ-1 couples. The Army PX (Post Exchange) was closing there to move to our area. Before moving, it had had an excellent sale on its merchandise to get rid of the inventory. I hoped to find some warm clothes because mine still had not arrived from the States, and my summery clothes were quite inadequate for the late autumn temperatures. The clothes inventory was quite well picked over, so later I had to go to Iwakuni for a coat and wool suit. But at Eta Jima, we chose a beautiful Noritake dinner set with coordinating tea set.

Our Army offered a course to learn the Japanese language, and a couple of the Australian wives came to take it. One said it was so difficult to understand the strange way Americans talk. Said she, "One American lady always said our country is pronounced 'O-strile-ya,' and actually it's pronounced `O-strile-ya.'" We Americans stole puzzled looks at each other.

After two or three lessons, the Japanese instructor offered to teach us some Japanese history. I thought that would be very interesting, as did several others. The course was quite brief: one lesson. He started out by saying, "The Emperor is descended from the Sun." Someone there must have told American Powers-That-Be because the Occupation did not permit that verbal opinion, and we saw him no more.

After a while, the husbands who were not out on missions decided to entertain us wives at a Japanese nightclub to see how much fun it was. We were escorted to the second floor balcony of a hotel where dining tables were all arranged for us. We sank down to the zabutons (floor pillows) furnished for seating. We didn't seem to be able to handle being on our knees for an extended time. Although we wives were wearing skirts, we discovered that through the use of some rather athletic gyrations it was possible to get our legs straight out under the tables. Then we had our first taste of sake (rice wine) and found it quite tasty and deceptively mild. Several pretty young Japanese ladies came around to pour it and then began playing hand games and singing. (Some thought they were geishas, but if so, they were ones called après-war geishas. They had not received the extensive training given the traditional ones.) What was not immediately apparent was that most of the entertaining was directed towards the men. Later photos showed husbands laughing as they joined in the entertainment — and wives not even smiling. There were no further similar parties.

Occasionally we went to the Officers' Club, now moved to the Post somewhere near the movie house, for a dinner dance. Japanese bands seemed to know "Dixie" was a favorite and that many people stood up during it. One Army officer had quite an attractive wife he obviously enjoyed adorning with furs, jewelry, and fine clothes. When she heard the opening notes of "Dixie," she was always the first one on her feet. We quietly called her Miss Southern Austria, her nationality. The dinners were pleasant social events, but the dancing segment seemed to send me back to adolescent wallflower rejection discomfort. Gary and I danced the beginning and last dances together, then he always likes to dance with other women we know. At first I thought that was nice and looked forward to dancing with the men. But here each man not dancing seemed to disappear while his wife was dancing. And so I sat alone, trying to appear entertained by watching the dancers. I became very insecure about my dancing and thought I must look extremely awkward while doing it.

We started to have some sort of routine when possible. A favorite was lazy Sunday mornings in bed, surrounded by the beautiful scenery outside our windows. We turned on the radio and Gary returned from the kitchen carrying a cutting board holding toaster, bread and butter, and orange juice. We first listened to Radio Free America news, occasionally getting a Russian broadcast, but after a few Sundays our preference became presentations on an Australian station by a singer named Anna Russell. She told us all about different operas, sang excerpts, and inserted her own interpretations. She was very funny.

Buying things became a new pleasure. The PX made its move to a few blocks from us. Having it close to where we lived certainly made shopping a lot easier. I found some lovely woolens there from the British Isles and a marvelously smooth and soft Egyptian cotton. The Commissary had been there all along, just beyond where the PX came to be. It carried excellent meat, and one of our favorite easy-to-fix meals was T-bone steak. It cost a whopping forty-five cents a pound, inexpensive even for that day! The same cut of meat later at Iwakuni cost sixty-nine cents a pound and was not nearly so tender nor tasty as the Army had. Could it have been because the Army was the supplier and took the best cuts off first? A little inter-service one-upmanship? An embarrassing incident once happened to me at the Nijimura Commissary on a Saturday when Gary was gone and the bank was closed. Our money was not American greenbacks. It was military scrip and looked like Monopoly money, which was what we all called it. In a singularly odd decision, five dollars and five cents were both blue. The five dollar bill was only slightly larger in measurement. I thought I had five dollars and chose some necessary items, only to find I had just five cents.

One evening Gary came home from Iwakuni with two large wooden boxes (how'd he get them there and from the boat landing to our house?) The Ship's Store had received a shipment of glassware, and tinkling could be heard when the boxes were moved. The price was excellent and broken pieces would be replaced. I believe it was something like ten sets of twelve different kinds of glasses.

We had a couple of less pleasing occurrences. One evening Gary was even more tired than usual and decided to go to bed rather than nap on the couch. By then it must have been spring because it was still slightly light when I retired, too. I was just drifting off in the early stages of sleep when I became aware the windows were rattling. "It must be windy," I muttered dreamily, but didn't recall that it had been. And then realization set in that the rattling seemed to be rather rhythmic, as though the windows were going to one side, then another. "Gary, Gary!" I screamed. "We're having an earthquake! You've got to wake up! We need to get outside!" There was no response at all from Gary, and I wasn't about to leave him there, but I did go stand in the bedroom doorway. He remained unconscious while I tried to slow my thudding heart. One afternoon sometime later we had another earthquake. The pictures rocked out from the wall, and I suddenly sat down hard on the couch, not because the quake was so strong, but because my shaking knees gave out. That one Gary felt in Iwakuni. The most frightening thing about earthquakes was recalling that one of the Army wives had said because we were between the mouths of two rivers, sometimes the water came up through the

ground in our yards. I pictured a movie I'd seen at some earlier time with the earth yawning immensely and people and buildings sliding and tumbling down into oblivion.

On one occasion we welcomed unusual entertainment on the Parade Ground: a group of Scotsmen, resplendent in their kilts, drums a-playing, and bagpipes a-skirling. Aye. LaVon was especially thrilled because they wore the plaids of her ancestral family.

Glen and Marilyn soon left, but we would see them again in Hawaii when we returned to the States. And we have seen them off and on through the years since they returned to our mutual old hometown after Glen left the Navy. Others of us began peeling off as housing became available at NAS Iwakuni. We moved there in April, into the apartment vacated by Margaret and Don Burdg. LaVon and Dean lived in the same building with Liz and Wells Rathbun in the unit between us.

Before we left Nijimura, we saw spring come to Japan and learned why Japanese families take off work to go to beautiful places to enjoy cherry blossoms in bloom. The ladies abandoned their somber winter garb and seemed to blossom themselves, wearing kimonos and obis of dazzling combinations of colors and designs. We strolled around the perimeter of our housing area, stopping under nearly every tree, resting on the sea wall, and enjoying the warmth of the sun. Then we moved to a new chapter, not nearly so scenic and calm, but certainly more convenient and homelike, with more attractive quarters, enabling us to see more of our spouses. It was particularly good to have all the VQ-1 families in the same place.

Nas Iwakuni

The especially desirable move came. Our unit would be in the first building on the left after coming in C Gate. I don't think there was anyone from VQ-1 on the end next to us until later when Norm and Rita Bull and Rita's son Tony moved in.

From the moment we walked in, I liked this place better than the one in Nijimura. The apartment was arranged similarly, but there just seemed to be a different feel to it. Although the walls were *the* green, they went well with the furniture. The fabric was upholstered with something like small needlepoint medallions, and having a pattern seemed to make them more homelike. Somehow the atmosphere was lighter, probably because the dining room had double windows at the end, giving us a straight shot of openness. It felt even cozier after I bought a bolt of white muslin and made café curtains. When living at Nijimura, we had begun a personal library with a subscription to Book-of-the-Month Club and Opera-of-the-Month Club, and after we'd been in Iwakuni awhile we bought an encyclopedia along with two other sets of reference books. Gary went to the woodworking shop and made three two-shelf bookcases for us. Little additions of things made by ourselves have always helped make us feel at home.

The apartment was slightly smaller, but it didn't seem to matter. Inaccurate memories say most days were cloudy in Nijimura and sunny in Iwakuni. We did miss our beautiful scenery from the bedroom. The kitchen was a more awkward arrangement with less cupboard space. But who wants to spend extra time there anyway, even with a stove that works? Some previous owner had put open shelves extending over the stove. Since I remembered all too well our Filipino geckoes running over dishes, those went into the cupboards with doors; staples went on to the open shelves. Then I made a curtain which covered at least those on the bottom shelf. That and the curtain on the window and door window also helped the kitchen look a little neater. No matter the shortcomings, we loved living there.

We soon became close social friends with two new VQ-1 couples in the next building, Shelby and Yvonne Bennett and Dave and Jody Hawbaker. The Bennetts had a son Gerry and Hawbakers a daughter Wendy. Other VQ-1 families lived varying distances from us but no one on the Base was too far away.

Gary used our small bedroom to build a hi-fi set, meticulously arranging small parts in the corrugated end of a piece of cardboard for easy access. He may have been inspired by Shelby, who built a massive speaker system after he and Gary went to Hong Kong for components. The day Shelby was to try it out, Yvonne, Jody, and I drove off the Base. When we returned we parked across the street and, with windows closed and before opening doors, we could actually feel the cannons and bells reverberating from "The 1812 Overture." We later learned that when Bennett's household goods were on their way to the States, that speaker was dropped with the rest of their shipment into Tokyo Bay.

How wonderful to have a life a little more "normal!" When not flying, Gary could come home for lunch. Summer came close on the heels of our April move. Some of us wives took morning swimming lessons in the immense Base pool. My memory says that pool was as wide as most pools are long and so its length then became proportionately longer. When we were done with our lesson, we jumped on our bicycles (I was now a much more accomplished rider than in my Philippine days) to go home in time to fix lunch and set the table. Our lunch was then refrigerated because Gary came home to change into his swimsuit. Off he and I cycled to the pool, swimming for twenty minutes or so, then rode back home where I put the finishing touches on lunch while Gary showered. If we finished early, we'd play at least part of a game of Cribbage in this 90-minute lunchtime. Often we'd swim again after the workday was done. Evenings we read or listened to the radio, eventually hearing songs by a new stateside sensation named Elvis Presley. The military disc jockey apparently was familiar with absolutely every US jazz and popular song performer, including their kinship with other people whose reputations were known only to him. He so often referred to "poignant" selections that we called him "the guy with the little poignant head."

Many evenings we'd also hear planes — our Squadron's and others (Gary could differentiate engine noises) practicing touch and goes. Their acceleration and deceleration were quite loud. When more and more Marines arrived, they favored us dependents by parking their planes with the rear ends toward our housing so we'd be sure to notice when they revved their engines, whether they took off or not. At least that's what it seemed like; I usually noticed the noise a lot more when Gary was gone and I was trying to remain absorbed in a book.

During that summer, I tutored a little boy in our place. He may have been eight or so, was very quiet and shy, but applied himself diligently to his reading. He was the son of the man who managed the Ship's Store.

It wasn't until we lived in Iwakuni that I discovered how wound up Gary was when he got off a flight. Besides the tensions about whatever he was doing, he had to have been affected by his three daily packs of cigarettes and copious cups of coffee while flying. He hadn't been around long enough in the P.I. for me to detect this, and at Nijimura usually had a day to unwind before his evening boat ride.

Sometime after we all moved to NAS Iwakuni, the skies opened, the sun shone, and a miracle happened: we were allowed not only into the VQ-1 hangar but also into the P4M. It was an Open House, and of course anything in it that was secret was closed off/covered up. But it was thrilling to see where our husbands spent so much of their time away from us. I did feel I was an intruder, though, and that I really wasn't supposed to be there.

Another event that had waited until our move to NAS Iwakuni was the obligatory call on the Commanding Officer and His Lady. The Halls lived close by, on the other side of our street and just around a curve that echoed the river. We were pleasantly invited to sit down on a loveseat bracketed by two trees in containers. In front of us was a coffee table holding a plate of appetizing-looking cookies. The Halls had a large Boxer dog that was roaming the room. Soon he strolled over to the table and stood slobbering over the cookies. Mrs. Hall strongly urged us each to have one. Reluctantly we gingerly picked up one each. About that time her attention was distracted, and quickly we thrust ours behind the trees. Fortunately she didn't ask us how we enjoyed them.

After awhile Yvonne and Jody, together with Dottie Flood, wanted a fourth for

Bridge and invited me to learn as we played. We spent many pleasant afternoons talking, laughing, and nibbling on Corn Curls. Although they played to win, we did it for fun.

Their teaching was quite thorough and serious, so when I felt fairly confident about my playing, I decided to go to Officers' Wives' Afternoon Bridge. I drew as my partner the Base icicle. She didn't smile. She didn't chat. Supposedly she had mastered Bridge in one year and because of her reputation, winning was extremely important to her. My palms should have become clammy. Trying not to panic, during the next thirty seconds I mentally reviewed in depth everything I'd learned. After we played a few hands, she got a terrific one, giving me a demand bid. I was sure I responded correctly, my longest suit since I had only one point. All I needed to do was keep answering with my long suit and secondary ones so she'd know where my strength lay. Right? She kept on bidding — and so did I — which led her to believe my hand also was excellent. Needless to say, we went down in flames. Desperately I apologized while avoiding looking directly at her, fully convinced I would have been turned to stone if we made eye contact. Aw-w-w, who needed O-Wives Bridge anyway? Our little afternoon group was lots more fun. Gary also learned Bridge and some evenings we played with a couple at a time, Bennetts, Floods, or Hawbakers.

There were many, many more occasions for socializing here than there had been in Nijimura. Most of our frequent cocktail hours/dinners were on the shore of the Inland Sea, at the Slipway Club on the other side of the runway. This was a positive way the guys could blow off pent-up stress. It was a much more private and intimate place than the regular O' Club, since we were usually the only group there. Other people on the Base really didn't quite know what to do about us. We were pleasant, they were pleasant, but there was not a whole lot of other interaction. It was just a lot easier to keep our exclusivity, and the parties were an excellent excuse to increase our wardrobes. Nobody wants to be seen in the same dress event after event! Several of us had the elegant and universally flattering Chinese-style brocade dresses made. They were perfect cocktail dresses — slenderizing, sleeveless, had high collars, and slits necessary to walk. The common joke was that any slits longer than ten to twelve inches meant you were a naughty girl. Since current skirt lengths were mid-calf, our reputations were not in danger.

When traveling toward the Slipway Club and Inland Sea, occasionally we caught a glimpse of one of our Squadron's P4M planes evading the smokestacks at the left end of the runway while circling the field practicing daytime touch-and-goes. As this very dark plane with its distinctive silhouette and no visible markings slowly and rather majestically moved, certain words would always occur to me: prowling, shark-like, predatory, and, in an odd twist, strangely comforting. I liked to think perhaps the last word was some sort of indication that we'd be watched over and kept safe.

Many of the activities at the larger and more eclectic O' Club were fun, too, and it was a good place to go for a frequent special dinner for the two of us, with couples, or accompanied by other wives whose husbands were gone. One of the latter occurred after some of them had decided I should learn to play golf. They were most encouraging, saying I was a "natural," and so we played on and on. The next day movement on my right side seemed a little limited, but it wasn't until dinnertime that its full extent was discovered. I ordered a steak and found to my dismay that it was impossible to cut, not because it wasn't tender — there simply was not enough strength in my right side. We all got quite a laugh out of that. I think I played one more time and decided that my stint as a golf natural was just a fluke — unless the object of the game was to see how high my score could be.

Sometimes we would go off-Base, Gary and I to explore, or "the girls" and I to shop in the area just outside the Main Gate. We might go even further away if someone had a car. For us women, our return to the Base was an interesting experience. If on bicycle, we were ordered to dismount and present identification. This was scrutinized at great length and compared with our physical being, possibly by-the-books the proper thing to do. However, we were absolutely certain that those Marines of an estimated maximum eighteen years old who imperiously looked down their noses at us from under their caps were acutely aware of every Round-Eye female on base. Occasionally some worldly twenty-year old apparently was able to realize we were not a grave security risk. As we returned by car, he gave a snappy salute to the officer he assumed was inside and waved us through the gate. When we passed, chortling with glee that we had thwarted him, he looked chagrined.

Occasionally Gary and I bicycled out to gorgeous Kentai Bridge, the famous one with five arches that created humps that made bicycling difficult for us and impossible for evil spirits to cross. Though still 1956, in this second viewing of Japanese springtime, again we were treated to a riot of color. Here the scenery was far different from that in Nijimura, but no less beautiful. Clumps of profusely dec-

Kentai Bridge in Iwakuni, Japan

orated cherry trees were in complete full bloom, as were the less colorful but nonetheless beautiful plum trees on the terraced hills. Yellow mustard added even more color to the terraces. In Nijimura, we had not seen how huge the crowds were of Japanese people taking off work and school. With the first spring there and in Iwakuni, plus a visit to Kyoto the next year before we returned to the States, we never tired of the metamorphoses. Winter had been somber and every view of buildings and roofs — everything visible — was black or at best gray, relieved only by a few off-white patches.

Not only did we have that famous bridge reasonably close (when I was teaching several of us teachers took our classes there at the same time as Emperor Hirohito's son was visiting), but we also could go to Miyajima, where we could see the often-photographed red tori. It is called "The Floating Torii," but only looks like it is floating when the tide is in. The tradition was for fishermen to go through it, supposedly to bring good luck. We took a train to the little town of Miyajima-guchi, about halfway between Iwakuni and Hiroshima, then a ferry across to the island. We loved the time we went in autumn — it was the only place we knew of where we could see Japanese Red Maple leaves, a marvelous spot of color against the omnipresent evergreen trees. One time in a grove of the maples with their brilliance accentuated even more by full sunlight, we came upon preparations for a picnic: tatami mats were laid out as the perimeter of a square. At each place was a napkin and box lunch of some sort. We feared, though, that once the picnic started, the orderliness would not last: each place also had a large bottle of Kirin or Asahi beer and one of sake. Elsewhere at water's edge near the tori, we saw a short play/dance given by a player with a happy facemask and another with a sad or scowling mask. Fortunately it wasn't so long as formal Japanese plays because the weather was quite cold, despite the sunshine.

It was also fun to go to Kinnan's, a Japanese restaurant very close to the Base outside the Main Gate. We could see into the kitchen and watch the cook make fried rice over a roaring iron range reminiscent of those of my mother's siblings on their farms. It was awe-inspiring: above a burner with its lid removed and flames licking upward to catch a rather large cast-iron skillet, the rice was tossed several feet into the air. Every grain appeared to be caught on its way down. This dish seemed to be something like an appetizer that came with the meal. However, often on Sunday evenings we'd call out to Kinnan's for a large order of it, delivered by a bicyclist carrying what looked like a wooden cigar box. Once the lid was removed, chopsticks could be broken off for use, if desired. At the restaurant, the next treat was having our own steaks cooked at the table on a hibachi. I think our steak dinners each cost about 300 yen. (Exchange was 360 yen = $1). This was a real bargain. A steak dinner at the Club was a whopping $1.35.

On the negative side, hygiene again was far different from ours and the lack of it was quite visible on our trips outside the Base. Gary's favorite slide from that time was of me against a display of the natural beauty. But in the distance, and unnoticed when he snapped the picture, was a man relieving himself. Another trip, a man was so entranced by my blonde hair that as I biked past he turned to watch me, all the while in full stream. With flaming face, I tried to ignore him. On more distant offsite trips, we were lucky indeed if we could find separate public bathrooms. Usually this happened only in heavily tourist-attractive sites. Mostly women's receptacles were urinals lying horizontally on the floor. One place, though, had separate stalls like we were used to. Each sheet of toilet paper on the roll was stamped in English with "Government Property." Did they want it back?

The Rainy Season seemed extra long that first spring in Iwakuni, lasting from April into July. Those lovely, sunny days we had experienced to view color became few and far between. We quickly learned that if we didn't go anywhere when it was raining, we didn't go anywhere at all. Arming ourselves with oversize Japanese umbrellas and plastic raincoats that nearly reached our feet, off we would go. One day in May or June a Japanese clerk at the Ship's Store told us that the thunder we were hearing meant the Rainy Season was over. She was wrong. It was July 4th when we took a boat trip to Shikoku, an island south of Honshu. It was quite foggy but otherwise a pleasant and welcome getaway. The most impressive feature of the trip was to Matsuyama, a feudal castle perched on a high hill. Soon after that trip the skies cleared.

We were treated to a couple of typhoons, or I should say the women and children were. The men and planes flew off to some safe area, usually the Philippines so they could take their swimsuits, or Okinawa where they had a house exclusively for Squadron use. While our facing the weather alone was daunting, at least we knew where they were, what they were doing, and that they probably would be back when the typhoon had disappeared. Since no one had a garage with their housing, families with cars were allowed to take them to empty hangars for shelter. I don't recall if dependent families who lived off-Base were allowed to do so, too. Actually, those two typhoons were kind of anticlimactic, considering the one we'd experienced in the Philippines. For the first one in Japan, I stuffed the very loose windows with towels to diminish their rattling and keep out blowing rain. But there was very little wind, just a lot of rain. The next typhoon had practically no rain, but it did have lots of wind. Looking out my upstairs window, I saw one of the Japanese workers on his motorcycle. Ve-e-ry slo-o-o-w-ly he came down the little hill from C-Gate and when he reached the fork where another street veered off toward the Commissary, his forward motion stopped and he was forced backwards down the street, out of my sight. It looked quite strange to see a motorcycle going backwards.

Several families of our Squadron chose to live off the Base. Most moved into a new development called Kurihara Heights, built by the enterprising Mr. Kurihara. The units offered perhaps more room and interior variety, but they looked a lot like packing boxes constructed of low grade plywood. For the most part, the Squadron people who lived there were more newly arrived than we, who now were becoming at least mid-timers.

Our pleasant life went on, and I even started becoming a little used to Gary's frequent absences, although I did warn myself it was never wise to become complacent. Unfortunately I was correct.

In August 1956, Gary, Karle Naggs, and two newer Lieutenant Commanders ("Mac" McIlmoil and Lee McHugh) in the Squadron, along with four enlisted crewmembers, were released from Squadron duties. The next day they would be returning to the States for a couple months to pick up and ferry back two new types of planes, jets designated A3D *Sky Warriors*. Frank Flood was to take Gary's place for the night's operational flight, and Gary was assigned the Officer of the Day Duty in his place. The mission that night involved two planes scheduled to go out an hour apart. The plane designated "62" was to be the first, but it had some minor problem. So Gary decided to delay the mission an hour. That meant the order of the two planes was switched, putting the other plane first and "62" second, another hour later. Gary stayed until "62" took off, not knowing he was the last person to see the ill-fated crew alive. Then he came home to get packed and prepare to leave early the next morning. A new man in the Squadron, acting as Assistant Duty Officer, relieved him to monitor the progress of the mission and the position reports. About 4:00 a.m. the phone rang. Gary took the call and said there was a problem; he'd have to go to the Squadron. I thought, "What can be so serious that they can't take care of it without him?" Several hours later he returned. I'd been pacing the floor because I thought he'd miss his flight to the States. Instead he said he needed some breakfast and I needed to go to Dottie's, Frank's plane had been shot down.

My stomach turned over. This sounded ominously close to my worst fears. And how could I help Dottie? What would I say? This was w-a-a-a-ay outside my experience. When I reached Dottie's, the house was full of Squadron women, all gathered around her. I think someone, perhaps not in the Squadron, had taken her little daughter Nancy out of the hubbub; I really never did know and never saw her again.

Everyone just kind of kept milling around, impossibly trying to find something helpful to do, all waiting for news. Once in a while one of the men would come by or phone, saying there was no news. After a bit, I noticed Bev Deane sitting alone at the other end of the living room, staring out the window. She and her husband Jim were newly married and had only been in the Squadron a little over a month. He was on the flight. Additionally, they lived off the Base, so not many people knew her. I may have met her at one of the Slipway parties but hadn't visited with her. Finally I decided she'd had enough "alone" time so went over and sat down with her. Trying to divert her mind from the situation a bit, I began asking about her life before Navy. She'd interrupted her studies leading to an M.D. so she could be with Jim during this overseas duty. We talked for quite a length of time about that and then what had happened earlier in the morning when the Squadron's Executive Officer and a chaplain came to her door, just like Jim had told her would occur if something happened to him.

The rest of the day has left my memory, but apparently by late afternoon we knew the plane was lost off the coast of China. It might have been that day or the next that the bodies of Jim Ponsford, a technical officer who lived across the street from us, and an enlisted crew member, Aviation Technician First Class Mattin, were found. That evening we went to Mattin's home to express our sympathies. He and his wife had three little children, one of them an infant.

The next morning I went to Dottie's again, as I'd told her I would. Approaching a couple of houses before hers, I could hear crying even with doors and windows closed. The front door was unlocked, so I went in and found her sitting on the floor in a bedroom, sobbing as she packed Frank's clothes. I put my arms around her and held her for as long as she wanted. I still cry when I remember that. And this whole incident is very difficult to write. For the most part it has been indelibly written in our brains.

What a tumult of emotions had to have consumed every one of us: shock, sorrow for those lost and their families, empathy with the families, unexpressed relief and guilt at feeling that relief, guilt at all the changes that had been made that meant someone else had experienced what perhaps might or would have happened to you, unexpressed anger at the situation and that it had to have happened to loved ones, and probably all sorts of other mixed-up thoughts.

Both Gary and I have lost the sense of time and how many days (not many) were involved, but it was decided that the men who were going Stateside to pick up the two new planes would escort the families back there. Talk about being between a rock and a hard place personally: I didn't want Gary to leave, but if he stayed then he'd have to go operational again and that was even more frightening than it had been before The Event. Either way I was in an emotional quagmire while being quite sure absolutely every other wife was, too.

The returning wives were very nervous that they would be interviewed by reporters, afraid they'd say something wrong (they weren't interviewed). In Hawaii, the four officers were questioned by the press, but could answer none of their questions. Apparently someone in an official capacity there handled whatever could be said. Gary can't remember a similar experience when they arrived in the States, and the families "escaped" as quickly as possible.

For many days after the families and the guys left, there was considerable activity to check what had happened — the plane had been shot down, even though this was technically peacetime. There were no survivors. There was very little publicity — we saw only one news item from Reuters in ***The Stars and Stripes***, and someone received a couple of clippings from the States. On a sunny weekend day a few officers took a boat and a wreath out into the Inland Sea to hold a little memorial service for Jim Ponsford. I was surprised only one other wife besides me went along. It was quite touching.

The whole experience is one that seems to resurface over these years more frequently than one would think. We corresponded with Bev Deane, the newly widowed bride, for a number of years, through her completion of Med School and internship. For a while she said that school did not help her cope with the loss as much as everyone else thought it would. Eventually she met someone, next married him, and after she said she was pregnant, we heard no more. While we were sad to hear no longer, we thought, "Good, she's making a new life."

And then we did hear from her again. Early in **1993** Gary's receptionist in Chicago took a call from some man who said he'd been in the Navy with a Dr. Grau who was a dentist and had lived in Huntley. Would this happen to be the same one? It was Dean North, and Gary immediately took the call. We'd lost contact with Dean and LaVon through some of their moves. Anyway, he said the former VW-3 Det A members, spearheaded by Stretch and Pauly, had had a reunion a couple years before and there would be another in June at Jim and Mary Jane Edixon's vacation home in Virginia Beach. Did we want to come? Of course we said yes! and spent a night first at the North's after several hours of wonderful reminiscences. We also learned that under the new Freedom of Information Act, some information about VW-3/VQ-1 had been declassified. Incredulous, we learned that reportedly there had been three survivors of the plane crash who made it to the China coast. One died immediately and two were hospitalized, then disappeared after release from the hospital. One seemed to answer the description of Jim Deane. We also heard Bev was digging for all the information she could. Not long after the Reunion, she called Gary, asking very specific questions about who was where in the plane, and who could have gotten out and how, etc.

Now flash-forward to late **2005**. One day we got a phone message from a lady who identified herself as Katie Shriver. She was Bev Deane's daughter, a ***Washington Post*** journalist, who was doing a special feature on the subject. Would we talk with her? Of course we would, and the upshot of all her research was most of an early May 2006 ***Post's Sunday Magazine*** with Bev and Jim's wedding picture on the front. Katie was stymied in her attempt to be able to close the door for her mother on this story that seems never to end, but she wrote an excellent article.

Back to 1956. Things returned to routine for most people again. With Dottie gone, we needed a fourth for Bridge, another one who played for fun. We found a newly arrived wife, Rita Fink. She lived in Kurihara Heights but was able to get into the Base quite easily, so for a while we were able to resume our frequent schedule.

I messed that up, though, finding a new experience to occupy me while Gary was gone. A teacher hired for a combined second/third grade class in the elementary school located behind our building had not shown up. Squadron wife Betty Rex had been teaching it for a few weeks, but she was a high school teacher and felt very uncomfortable with this age group. So I filled out the application, seven copies to make by hand in those pre-computer days since my typewriter could not come up with legible copies using carbon paper, and got the job.

The school was close to our back door and across the street from the Commissary, making them both very convenient. Among my second-grade students were three very intelligent, over-achieving daughters of fellow Squadron families: Joe and Charlene Schmidt's Linda, Fred and Marion Groenert's Dee Dee, and Wells and Liz Rathbun's Betsy. Having a combination class was quite overwhelming. I was to teach advanced second graders and slower third graders, theoretically, according to the powers-that-be, making them about the same grade level. That was not true; most of the second graders were more advanced than the third graders. It was difficult working in all the necessary subjects because I felt I had to have many more small groups. I may have had four reading groups, or I may have had two or three for each grade, and I know I had more than one mathematics class. It was very challenging, and when leaving to return to the States I felt I probably should write our friends a letter of apology. This feeling of inadequate teaching was somewhat relieved many years later when I learned that Linda Schmidt had become a teacher herself and remembered me as her favorite. Her mother Charlene sent me the class picture taken outside our classroom. The first room on the left of the entrance, it was a lovely room. One wall was all glass, and most of it slid back as a door to open during pleasant weather.

2009: That building has been replaced with a new school. About twelve years ago, the daughter of a friend of mine was teaching in Iwakuni. Her parents visited her, brought back a photo, and from the looks of the photo, I believe it may be located where our residence building was. I chuckled to myself, knowing it's now a Marine base and remembering when General Pace became the new head of the Marine Corps. Growling, "You're a fighting group - if we wanted you to have a wife, we'd have issued one!" he sent all the Marine dependents back home.

There were a couple of memorable experiences during this 1956-7 school stint, one of which will be related here and the other a bit later on. This one involved parent-teacher conferences. Report cards had been sent out just before them. Liz and Wells Rathbun came, and I knew it would be a fascinating experience because of the markings I'd given Betsy. But it seemed like the only way it could be done, as long as there was the opportunity to explain verbally. Wells came blustering in, wildly waving the report card and explosively shouting, "How can you say she isn't working up to capacity when you've given her all A's!" To which I responded that she was capable of more advanced work than I was able to give her. She was far and away above second grade in her ability, but my own capabilities, given the hours and overload, made it just impossible. This seemed to placate Betsy's parents, especially when I suggested that when their geographical situation changed to Stateside, they should try to get her into a school that could challenge her adequately.

Wells and Liz were an interesting couple. Occasionally they would have a dinner party for Squadron people, serving excellent food. But none of us could understand how we could all squeeze into their abundantly furnished housing unit. They

Newly-arrived EA-3 Skywarrior, mid-1950s. Official Navy photo.

liked to buy three of everything exotic: one for them, one for Aunt Somebody, and who knew the other. Eventually they had to build a little house in the back yard to hold everything the apartment wouldn't. All it lacked was a quarter moon to make it look like an outhouse. Even though we tried to time when we left our place (as did Dean and LaVon), it was impossible to get away on every occasion without meeting Wells. One time he saw us and excitedly said, "You've got to come in and see what we just got!" Obediently we complied, and he jumped up and down on a brand new Japanese tea table to show us how marvelously strong it was. Things like that weren't so bad, just a little strange, but he knew neither the North's nor I drank alcoholic beverages. He'd verbally grab hold of us, saying, "You've just *got* to taste this new drink I have!" When we politely tried to decline, he'd say with almost a threatening demeanor, "I consider it a personal insult if I offer a drink and it's turned down!"

This happened several times until one day Dean got hold of him outside, saying, "Wells! Wells! We just got a new blender! You've just *got* to try this terrific new drink we made in it!" Wells uncomfortably tried to demur, but Dean kept on with the same dogged tenacity previously exhibited by Wells. I think it cured him, at least as far as four of his neighbors were concerned.

Finally the day came when Gary was to return from his Stateside adventure. I think I taught in the morning, then took the afternoon off since he was to arrive later on. I was too nervous to be able to concentrate on my duties. Most of the Squadron and many of the Base people went to see these two impressive new jets come in, the first of their kind flown across the Pacific. I stood on a slight rise, about the height of a table. One of the planes came roaring along in a fly-by speed run at about eye-level. It was thrilling — and chilling. It circled out over the edge of the Inland Sea, then came in, dragging a parachute behind it. In a flourish it folded up the tail and a wing, since it could be carrier-based as well as land-based. I thought it a little odd, just folding one wing and the tail, but what did I know? Soon the other plane also landed, and Gary was out of the first one. I was again in his arms in one of those reunion embraces so eloquently shown in newsreels, TV, and movies that always brings me to tears. The next morning when he went to the Squadron, one of the maintenance men called him over to the plane. There was only one deadbolt holding on the wing that hadn't folded up on arrival. They had flown from Guam with that limited wing support! And he wondered why I worried when he was flying?

Eventually we got home from the Welcome events, and Gary sat on the bed to take off shoes to change for dinner at the Club. Hmmm. What was that tiny jingling sound? We stopped talking, and he moved again. It *was* bells! Some invaders had been in our apartment while I was out of it! We decided we'd better check the sheets and sure enough, we'd been short-sheeted. The obvious suspects, who vigorously swore to their innocence were Jody and Yvonne. They had been let in by the maid while I was at school in the morning. We got revenge though. About a month later, Shelby returned from *his* trip back to the States to pick up another P4M. Jody and I got into their apartment while Yvonne was gone. We set several alarm clocks at different times and craftily hid them here and there throughout the bedroom and closet. The Bennett's didn't seem so suspicious as Gary and I had, so they didn't discover any clocks until they went off at various times during the night. They didn't think the situation was nearly so funny as we did.

As a result of the long separations and in the long-time tradition of the Navy, both Yvonne and I were pregnant at the same time, she due about a month later. By this time Jody and Dave had had their baby boy David, who was called "Dabit" by two-year old Wendy. I was fortunate with my maternity clothes: I designed and either made myself or had a seamstress make most of them from raw silk. At my mother-in-law's invitation, I sent her samples and she made me three or four tops to coordinate with the silks.

This Christmas was much more pleasant than our first one, helped no little bit by a sturdier tree in our homier atmosphere. A New Year's Eve party at the O' Club seemed a wonderful way to top off our full year in Japan. While dancing at 12:01, some of us whispered, "This is the year we get out of the Navy." While we were all finding this an incredible and unforgettable experience, we were eager to start our new lives.

A few months later, the other school incident occurred. Although I'd managed to handle all-day sickness, not just mornings, I was over that phase. However, one day I began feeling quite ill and desperately needed to use the restroom. Leaving the room in charge of one of the responsible kids, I hurriedly left, then asked the Office to get a substitute for the rest of the day. Returning to the classroom, I found all the kids sitting quietly and alertly at their desks, hands folded in an orderly manner in front of them, smiling brightly — and knowingly. Said one child, slyly, "Are you going to get your baby, Mrs. Grau?" Enough of them had younger siblings that apparently they were the "experts" who instructed the others in the progress of pregnancy. It was so sweet; I could have hugged every one of them!

Since I was teaching, we had hired a maid again. It wasn't too long before her pregnancy outdistanced mine. She was to eat lunch at our house while at work, but it seemed like our food supply dwindled a lot more rapidly than I would have assumed. One weekend when Gary was gone, I had my breakfast without really examining the refrigerator contents. The Commissary closed at noon on Saturdays, and I would have my lunch after that time. Great was my dismay to discover that

Ground Zero, Hiroshima, Japan

not only was there nothing for lunch, there also was nothing for dinner, nor any Sunday meals. I coped somehow, probably with the help of my friends and dinner at the O' Club. Not long after this experience, Gary came home about nine o'clock one morning and found her filling to nearly overflowing our largest frying pan with meat, eggs, and several other side items. He fired her. If she'd been a good maid, we might have tried reasoning with her, but the house never looked like she'd done any cleaning. Her time seemed to have been filled with eating.

Now spring was again approaching, and we didn't have too much more time in Japan. We and the Bennett's decided to go to Hiroshima to visit the museum devoted to the first atomic bomb blast. It was quite an unnerving experience. We, clearly Americans and the only ones present, went into the museum. It grew very still in there. The only sound was the click, click, click of the high heels Yvonne and I were wearing. We occasionally whispered to each other, but soon left. Outside we walked toward Ground Zero, the American Exposition Center ruin, passing a structure somewhat resembling the top of a covered wagon. We understood its interior was covered with names of those lost during the blast. A lone Japanese military veteran stood outside, looking accusingly at us.

The grounds had been made into a park, but the trees, still without springtime leaves, were creepy. They were all the same height and had weird protruding knobs that I thought looked like they might be tumors. We walked a little farther, then turned around to go back to the parking lot. Near the name memorial, there were now perhaps a hundred military veterans, all silently looking accusingly at us, all wearing the white pajama-type clothes that designated them, some with sleeves or pant legs removed to display the fact that they'd lost an arm or leg. It was not a comfortable place to be. As Mark Twain so eloquently understated when he thought he saw a ghost one night, "We went from that place."

In April/May Gary took leave. We didn't know if I'd be returning Stateside alone in May or if we'd be staying in Japan until after our baby was born, due in August. Anyway, we wanted to see more of this beautiful country that was presently our home. John and Rita Fink, she now also pregnant, braved Japanese train travel and went with us to Kyoto and Nara at this lovely time of year before the rainy season began.

One of the first things we saw were streams of water with bright colors running in them. We were in the area where fabrics were dyed and then washed in the streams to set the colors. The hanging cherry trees supported on platforms were also a new sight, some platforms extending quite far over pond waters and therefore reflected, making a doubly spectacular vision. They were a deeper pink than the Kentai Bridge variety. We visited numerous shrines and buildings in Kyoto and Nara, most spectacular of which was the Gold Pavilion. (In 2006, our son, who is an airline pilot, had a flight to Osaka and took along his wife who was born in Japan when her Air Force father was stationed there. Spending a couple of extra days, they took a side trip to Nara, and we laughed when we saw their photos. They're nearly identical with our slides of the huge Buddhist shrine there.)

Hirohito's white horse was gorgeous, stabled at Nara's Deer Park. Japanese gardens evoke great serenity, and Kyoto's Imperial Gardens certainly outdid themselves. Another favorite site for me was the Silk Mansion where I purchased a couple of pieces of beautiful silk. Lovely also were the products of some "factories," actually just small shops, where Satsuma ware, cloisonné, and damascene were manufactured. We bought examples of each and also found a kakemono ("hanging thing:" a scroll) which eventually ended up in a narrow tokonomo ("a place to hang it") which we had constructed when we built our house in the country. On a narrow shelf under our kakemono, I liked to put a small red lacquer bowl. All these are among my most prized artifacts, along with a few items Gary brought back from Okinawa and Taiwan.

When we arrived back in Iwakuni we found Gary's orders waiting, and glory be! He would be returning to the States with me. Karle Naggs, who'd been on the same set of orders to VW-3 Det A in 1955, was also to go to VU-7 at Brown Field near San Diego. Jack Reynolds did a nice favor to get them that assignment. We'd decided I'd go home to my parents for the summer. Gary's discharge would be August 21st, supposedly twelve days after our baby's birth. Gary wanted to complete his undergraduate work where we met, Cornell College. That's what seemed to be our true home, since both sets of parents had moved after our high school days.

I resigned my teaching job and handed it over to Rita. One morning a couple of the wives came to the door rather early. They told me we were going to gather the other wives for a come-as-you-are-going-away breakfast for me. This was fun! We walked hither and yon across yards, many of us in pajamas, robes, and slippers. Some of the non-Squadron residents rushed out of their homes, very concerned because they thought we'd lost another plane. We eventually ended up at Charlene Schmidt's, and I still treasure photos of the event. My going-away gifts, most of which I still have, came wrapped in small individual cereal boxes. In each was a different pair of earrings.

Not many days later I was again the guest of honor, this time at a surprise baby shower billed as a tea given by the Executive Officer's wife, Lorene Struthers. Once more I received some lovely gifts. But the most amazing decorative item there was a large Japanese umbrella from which the paper cover had been removed and each rib covered with real flowers. It was truly unique and must have taken hours to achieve.

We had a joint going-away party with Karle and Gary as honorees at the Slipway Club. The joke gift for Gary was a rolled-up mattress since he was known as Horizontal Grau — he could sleep anywhere, any way, and any time. The other joke gift was a Swiss Army knife with all the gadgets on it. This was billed as a do-it-yourself dental kit. The lovely serious gift was a wooden tray carved with the Squadron emblem, VW-3 Det A and VQ-1, and our names, still a treasure and on proud display in our home.

We had to arrange in advance a long-distance call to Gary's parents in Chicago. They had indicated they would like to meet us in San Francisco and needed the date. We gave them minimum information because the call cost $17.00 a minute! I don't remember where on Base we had to go to arrange and make this kind of call.

Our departure day came, and Gary deposited me where we were to catch the plane to Tokyo. He decided he needed one more time to tell everyone good-bye at the Squadron. Some work was being done there, and to depart from the hangar, he had to crawl out the Supply Office window, his former and most recent additional job. He came face to face for the first time with a high school classmate who'd been there at Iwakuni nearly all the time we were. Their brief visit made us just about miss our plane.

Once we'd confirmed the arrangements for our flight to the US and checked into a hotel in Tokyo, we set off for Frank Lloyd Wright's famous Imperial Hotel for dinner. Then we returned to our room to figure out what sights we wanted to see the next day. At four a.m. phone call changed that; we were to take off from Haneda Airport early that morning. The departure schedule didn't give us any time to dawdle. We hustled around the room, scooping up our few belongings and stuffing them helter-skelter into one bag or another. The time change going that direction, and its being made much more quickly than the one hour per day on the westward-heading ship from the States, made it more difficult to know what the accustomed hour really was. In what seemed to be quite a brief time, we had our evening dinner on still another famed island of World War II, Wake Island. And then it was dark. We sat with a man with unusual wings on his chest. Later I asked Gary what they were. He replied, "He's a submarine officer. Boy, I sure wouldn't want that duty! It's really dangerous!"

Our next stop was Hawaii, where we would have a fifteen-hour layover. The Carlstroms were there and let us use their Jaguar sports car to take a little tour of the island. Getting into the seat itself was quite an experience for me, nearly seven months pregnant. Commander and Mrs. Hall joined us for dinner and then we were off again, this time for good old USA. (She didn't say anything about the cookies.)

I was awake most of the night. Sitting backwards on the wing, I could see the red glow of an engine and stared at it, awaiting the flames I was sure would break out at any time. In front of us was a crying baby. When it finally fell asleep, its inexperienced Japanese mother decided she should change its diaper, which awakened it to the next round of crying.

Finally we literally hit the ground at Travis AFB, bouncing several times from about twenty feet up. Gary got an extremely dirty look when the crew left before us and as they passed he said, "Can I log one of those landings?" When we left the plane, the air was the freshest ever smelled! We'd gotten used to the aroma of rice paddies with their human fertilizer and could scarcely remember actual fresh air. While Gary checked in, I staggered to a complimentary bed for a few minutes; then we climbed on a bus for a two-hour trip. I fell asleep before the engine was started.

Gary's folks weren't due for about twenty-four hours, so we spent the next day doing some sight-seeing, then decided we'd go to the brand new Cinemascope movie. One segment of it was of the marvelous spring in Kyoto and Nara, where we'd just been. Afterwards we walked into Union Square, and across the corner from us was a Scandinavian couple who'd been on several tourist buses with us in those places. The world is indeed small!

The happy reunion took place. We headed off to see the Stockton relatives, then south to see more kinfolk and for Gary to report to Brown Field where he'd be the next three months. I took off for Omaha and the joy of being with my family again for those same three months.

Anxiety was not over yet: we definitely were counting on Uncle Sam's paying for the birth of our baby, due August 9th. As each day passed, we grew more anxious. Finally at 3:15 a.m. August 20th he arrived — the day before Gary left active duty. Now our different new life could happen, filled with old and new (our dear college friends), unknown events, and a new baby boy.

In retrospect, and at that time, I truly feel that we wives served our country, too. We had faith in our husbands, believing that they were performing a valuable and very dangerous task — whatever it happened to be - and giving them some sense of normality in an otherwise extremely stressful life. I don't remember ever trying to get Gary to tell me what he was doing and presume most other wives behaved in the same way. Those two years were probably the most important of our life together, setting a pattern of independence, interdependence, and confidence in each other, plus giving us some wonderful lifetime friends. Despite my loneliness and the negative aspects of life in the Philippines, I still look back at being there with fondness. I would have liked to see more of that land where my adult life really began. And despite all my tears of fear, I believe in my lifetime thus far there have been far more tears of joy than sorrow, more moments of enjoyment than displeasure, and in instances of situations I can do nothing about, I try to follow that advice given my children: "If you can't change an intolerable circumstance, then you must change your attitude."

Afterwards

It's amazing how many of the reminiscences are so very clear fifty years after their happening. And it's not surprising that there are many not so clear. We've had at least annual contact with many fellow Squadron members and their mates through Christmas letters and some occasional sightings. There have also been unexpected reminders, a few of which were in the body of this narrative. Contacts I can remember:

1957: Gary drove from Brown Field to Fort Dodge in the Volkswagen we ordered and had delivered in Japan. It arrived three days before we were to leave because the Suez Canal was closed, and the ship carrying our little Bug had had to travel around the Cape of Good Hope at Africa's tip. When we received it from the Tokyo dealer, we rode around as much as possible in the area near Iwakuni to see if we'd missed anything. Then it was put on another ship and delivered to San Francisco where Uncle Hugo picked it up for him. We drove it for seven years, and due to all the exposure to salt air, much of the Polar Silver paint peeled off. When we took it to be serviced the first time, the German mechanics lifted the hood of the engine and started laughing. The Japanese dealer had inserted a metal label written in Japanese, and thereafter the mechanics always called it the Volkswagen "Made in Japan."

On his way back to Iowa, Gary spent the first night in Las Vegas and had an unexpected — and totally pleasing — several hours visiting with Betty Grable! The next night he spent another pleasing one in Salt Lake City with Dean and LaVon.

A few days after he got to meet his new son in Fort Dodge, we returned to our beloved Cornell of Iowa college town so he could complete his undergraduate work. As he was walking toward campus the very first day of classes, a car pulled up with a couple of guys who asked if he'd like a ride. He got in and soon found out the driver, who had been a freshman my senior year, had just completed two years in the Navy. They didn't know each other because Gary was gone from school by that time, but they discovered they were in the same fraternity. The driver's Navy duty had been as a radioman on a ship named the USS Manchester, operating in the South China Sea. Gary said, "Were you "Old Bet" or "By Gosh?" Amazingly, they decided they had probably talked to each other several times.

1958: With Gary off from school for the summer, we decided to take a trip east to spend a weekend with the Hawbakers in Pittsburgh. We met their families and went sight-seeing, all for a wonderful time. One of the funniest things we did, though, was call the Bennett's in Tombstone, Arizona. We didn't have a phone number and, in any case, we had to go through a telephone operator who was located in Bisbee. There was no listing for Shelby Bennett. One of us remembered they were presently living with Yvonne's mother, whose name we didn't know. But we did think she was a teacher, and lived next to the school. The operator knew who we meant! Mission accomplished.

1959: After graduation from Cornell, Gary entered Northwestern University Dental School in Chicago. In conversation with a fellow student a year ahead of him, he learned the man had flown in the search mission to find our downed plane.

1960: It was my tenth high school reunion in Spencer, Iowa, where Glen and Marilyn Carlstrom were living, so we made arrangements to see them at that time. Subsequently we had visits several other times when in that general geographic area to visit family or for other reunions.

1964: We added a daughter to our family. After dental school graduation and a year's teaching at Northwestern, Gary and I visited a dental office he was considering purchasing in Barrington, Illinois. In casual conversation, we commented on the several oriental decorative accents. The dentist, Dr. Ron Fabrick, said as a dentist he'd been stationed in Iwakuni, Japan, until 1956. He and his family lived at Kurihara Heights and probably left about the time we moved there from Nijimura.

In the'70's: Rita Fink said she was coming through Chicago with a rather lengthy layover at O'Hare, so I drove in to meet her for lunch there. Later we went to an American Dental Association Annual Meeting at Disneyland and made arrangements for John and Rita, along with their younger son, to come to see us. On another trip to California, we were invited to their home for dinner.

1980: We hosted a VQ-1 mini-reunion for the "young" Squadron people at our country place fifty miles northwest of Chicago. Shelby, Yvonne, and their youngest, daughter Shelly Bennett, drove from Tombstone; Dave and Jody Hawbaker flew from Pittsburgh; and Karle and Alice Naggs with their two children

Flight pattern: Dave, Shelby, Karle, and Gary at our place in 1994.

drove from Wisconsin. Shelly drove the kids to the County Fair while the adults talked and talked and talked. The kids seemed to get along as well as the adults did!

1981: Our son was at Luke AFB, training in F-15's, and we decided to visit over July 4th so we could see our year-old grandson. We contacted the Bennetts who were planning a trip to Phoenix to buy a truck for son Gerry. He'd been about the same age as our grandson when we last saw him. It was a great visit.

1993: As said before, we attended our first VW-3/VQ-1 reunion at Jim and Mary Jane Edixon's Virginia Beach vacation home, all staying in houses around theirs. What fun! Bob Galloway gave each guy an artist's rendition of the P4M. He had commissioned it, and all signed their names on each other's treasured picture.

1994: We had another VQ-1 young guys' reunion, this time at our home in Schaumburg, Illinois. Present were Dave and Jody Hawbaker, Joe and Charlene Schmidt, Bob and Nan Galloway, and Karle and Alice Naggs. A year or so later, Dean and LaVon North stayed overnight with us on their way to Dean's high school class reunion in the far northland.

Mid-1990: Another well-attended wonderful VW-3/VQ-1 reunion was at Jack and Ursula Gillooley's summer home in Tennessee's spectacular Great Smoky Mountains.

Late 1990's: Still another VW3/VQ-1 reunion took place, this time at Andy and Irma Kruzich's near Seattle and its gorgeous views. This was the most recent of those reunions we have been able to attend, unfortunately.

A Filipino gentleman set up a mailing service business in our neighborhood. About every other year, he went back to the P.I. to see his family. He told us the former NAS Sangley Field is now a luxury resort with high-rise waterfront buildings the length of the former landing strip. Who could have foreseen the change from Quonset huts!

Early this century, Glen Carlstrom passed away and we attended his funeral in Emmetsburg, Iowa. In what can only be described as a truly surreal experience, I felt quite off balance! A panel of photos of Glen's life before Marilyn contained several that were duplicates of those in my Spencer High School yearbook. In the next panel appeared a photo from their Philippines wedding with my husband holding a sword over their heads. When we were seated for the funeral in the church next to Bob Galloway and his wife Nan, someone poked me and asked if I was Nancy Porter. It was a classmate of Glen's and I'd been in her court when she was the county Dairy Queen. Later seated with Galloway's, we were having a meal. Seated catawampus across from me was another of Glen's classmates whose father owned the furniture store next to the shoe store my father managed. His wife, not a Spencer native, sat next to me, and we chatted about their son who went to our college at the same time as our son. This was really an intermingling of several different chapters, if not lifetimes.

2004: Dave Hawbaker, whose wife Jody had died a couple of years before, came through Minnesota on a motor-home bus trip. We had dinner with him two evenings and did lots of talking.

2005: Gary's family reunion in Iowa blended the families of twins from several generations ago. Conversation with three previously unknown "cousins" revealed that all three were connected with VQ-1, one a former pilot, one a former mechanic, and one whose son was currently a mechanic.

2007: Our daughter moved to California, not too far from Pauly Varner. We traveled to Pauly's, went out for lunch, and viewed her gorgeous roses. A couple of months later, we were in Iowa and drove back to Minnesota through Emmetsburg so we could catch up with Marilyn Carlstrom.

2008: We planned to attend another VW-3/VQ-1 reunion, this one hosted by Don and Margaret Burdg in Oregon. Whoopee! The memorable events continue with these wonderful people! Unfortunately, we were unable to do so, another disappointment.

2009, October: We attended a program for the Honor Flight and after it, in conversation with the speaker, he mentioned a relative of his who flies the F-18 was just ordered to Iwakuni, Japan. The place is much changed, with runway lengthened and extensive building. It is now known as the most impressive Base in the Pacific.

And the beat goes on . . . . .

Nancy Lee Grau

Autobiography

April 22, 2011

Nancy Grau was born in Iowa and lived there until her Far East adventure in 1955-57. In 1954, she graduated from Cornell College, Mount Vernon, Iowa, where she had met her future husband Gary.

Nancy taught elementary grades in Iowa, Illinois, and Japan. She also taught Jr. High English and was a K-8 Learning Center Director in Illinois. Additionally, she was Executive Secretary for a firm on Chicago's Michigan Avenue and managed both Gary's general dentistry and orthodontic front desk offices over a number of years.

They have two children, Gary Wayne, Jr., born 1957, and Leslie Anne, born 1964. Wayne has a son and two daughters as well as two granddaughters. Leslie has one daughter.

Nancy has many avocations. The principal ones are needlework and American history. This latter interest was heightened by twenty years of volunteer work in the McHenry County, IL, Historical Library where she cross-indexed extensive biographical files, plus the opportunity for frequent visits to Springfield, IL, for detailed research into Illinois history. She has presented three papers at annual Illinois History Seminars. Additionally, as part of county Sesquicentennial celebrations she wrote a monologue and a biography of the man whom the county was named. This book was placed in a Time Capsule to be opened in 2035. She currently researches Daniel Pope Cook, for whom Cook County, IL, (where Chicago is located) was named.

She now lives in Minnesota in an Active Senior Community where she again does volunteer work, principally managing two libraries.

Marilyn Carlstrom
April 6, 2011

MY GREATEST ADVENTURE
The Beginning

It was November of 1954. The engines of the double-decker Boeing Stratocruiser droned overhead as the plane winged its way across the Pacific toward the Philippines. Staring wide-eyed out the lounge window, I could see what the steward was pointing to far off in the distance. It was the coastline of RUSSIA!

My mother Edith and civil engineer stepfather Jack were living in the Philippines where Jack was overseeing final construction of an American sponsored hospital designed to repay the Filipinos for their efforts in supporting the Americans during WWII. Happily, for me, they invited me to come over for a visit before they had to return to the U.S. At the time, I was a receptionist for Shell Development in Oakland, California, a job I enjoyed but which stood no comparison to the opportunity I was offered to experience a different part of the world. So I quit my job, packed my things, and suddenly found myself in very unfamiliar surroundings and a long, long way from home. That breath-taking glimpse of Russia was after long, cold hours the previous night sitting on the tarmac in Attu, Alaska, shivering under blankets because the heat had to be turned off while the plane took on fuel for the next leg of the trip. That leg took us to Guam, where the next evening we passengers again waited patiently in a Quonset hut for refueling while 'enjoying' the heat made even more oppressive by a typhoon that had just passed over the island. We proceeded on a roller coaster ride at the outer limits of that typhoon most of the way to Japan, our last stop before landing in Manila.

During that stopover, I opted to deplane in order to pick up a few toiletries that I thought were important, so I headed for a kiosk to buy them. Passing through the airport, it struck me that no one looked like me; almost everyone was quite a bit shorter and had beautiful glossy jet-black hair. At the counter, I picked out what I wanted and held out my cash. To my utter astonishment, the salesgirl smiled politely but shook her head and would not accept my money. She then pointed out — luckily for me in understandable English — that I had unacceptable dollars instead of yen, and thus I experienced my first ever culture shock, we Americans were not the only people in the world.

Boeing Stratocruiser. Based on the B-29 bomber, Boeing converted wartime technology to luxury civilian air travel.

My parents lived in Quezon City, a more affluent part of Manila, in an attractive house designed by a local architect for himself and his Danish bride-to-be. When that union didn't happen, Carlos rented the residence to my family. Typical of many 'nicer' Philippine homes, it had airy spacious rooms, a completely modern kitchen, and a small area off the kitchen for the live-in help. Our live-in was eighteen-year-old Florencio. I cannot but smile when I remember him. He had such a happy, generous spirit, hardly representative of the gruesome experiences he had lived as a younger child. Being from Samar during the Japanese occupation, he had witnessed his uncle having to dig graves for himself and other adult family members before they were executed. His gruesome history however, didn't seem to dampen his positive outlook. I don't remember how he ended up in Quezon City, but we were thankful he did for his upbeat presence around the house was a joy for us all. As the houseboy, one of his main jobs was to frequently wax the considerable amount of tiled floors. This was done by placing wax under the husk of a dried-out coconut under the arch of the foot and then sliding the foot forward and back. It was simple but effective, and a really good workout. Once in a while he would call, "Come on, Miss Marleen," his way of pronouncing Marilyn, "let's affly the wax," and both of us would scuff and polish away, then stand back and proudly admire our gleaming floors.

Our maid Angelina came in each day to cook and do some light housework. She was a jewel of a person who had a quiet way and loved to laugh at a good joke. She was married to Buon, an enlisted man in the Manila constabulary, the local police force. Unlike Florencio's tragic experience, during the occupation she had been able to find work and lived a somewhat normal life in spite of the constant presence of Japanese soldiers. She was a very important part of our daily lives, and it was primarily through her that we all learned much of Philippine culture and ways. I once lost my voice and her native home remedy was to cut three calamansi (a local lime-like fruit) in half, press them against my throat, wrap with a cloth, and have me sleep that way for three nights. It was unpleasant to say the least, and I still don't know if that was the miracle cure or not, but I did recover.

Angelina was also a good cook, taking particular pride in preparing native dishes. On occasion, she would stir up a big pot of adobo, a hearty pork stew with vegetables that was a particular family favorite, or make lumpia, the Filipino version of an egg roll, for a special treat. One of her daily duties was to mix up a dog food concoction from her homemade recipe with supplies she bought at the open market and served it to my parents' cocker spaniels. Occasionally I would accompany her to that local market to buy mung beans or other special vegetables not found at the Embassy commissary. Unlike the antiseptic orderliness of our American food stores, this was a raucous place with unpleasant pungent odors, hawkers aggressively selling all sorts of exotic produce – and flies, hordes of flies. I loved the atmosphere and the excitement. It was amazing, however, to think that people actually survived after eating the chickens that had been hanging around for hours in the heat.

Civilian Americans in Manila were under the aegis of the American Embassy and were a small, cohesive group that spent much of their free time seeing the sights and otherwise enjoying each other at social events. One couple became close friends with me. I believe as a surrogate for their own daughter who stayed in the States to attend college. Because my parents were not partygoers, the Smith's often asked me to accompany them on their evenings around town. At the Officers' Club, one evening a handsome young naval officer introduced himself as Glen Carlstrom and asked me to dance. This was surprising, considering he was in the company of another very attractive young lady. Things had gotten off to a rather 'sticky' start since the young lady was irritated with me for accidentally spilling a drink on her dress. After my apology, 'Oh that's all-right, it will only stain' was my dismissal. I think the young officer felt sorry for me, and so he asked me to dance. I stepped into his arms — and began the greatest adventure of my life.

Dating someone new is an adventure under the most normal of circumstances, but dating long-distance and in a foreign country was even more of a challenge for Glen. Luckily, he had a little black Ford that made countless long trips from Sangley Point around Manila Bay to our house to pick me up. Our first date was a road trip to Lake Taal, a beautiful site with an inactive volcano about 30 miles from Manila, where we spent a leisurely day having lunch in the Tagaytay Hotel and enjoying lots of 'getting to know you' conversation. I learned that he hailed from Spencer, a small town in northwest Iowa where he lived with his mother, father, and a younger brother, and that his father owned and operated a petroleum distributorship there. (At that time, my only knowledge of anything in Iowa was from reading portions of the ***Des Moines Register*** that my California high school News teacher touted as a sterling example of straightforward responsible reporting. A foreshadow of my future life in the common sense, no nonsense Midwest.) When we returned to my home that evening, I wasn't certain the attraction I felt was mutual, so I was pleasantly surprised when he called me the next day for a second date.

From then on, he was my 'steady' and my tour guide for anything I hadn't already seen in Manila. We visited some of the usual tourist destinations like Malacaynang, the presidential palace, with its impressive chandeliered great ballroom and strange looking native deer that roamed the grounds and Santo Tomas where political prisoners were forced to survive unbelievably harsh conditions. One evening we went for a stroll on the Luneta, Manila's largest and most popular public park. It was one of the only places I recall being where vendors sold the balut, a hardboiled seventeen-day-old duck egg, complete with embryo still intact. Glen challenged me to try one, but the idea of that 'delicacy' no matter how tasty it was to the Filipinos, was not for me. We also found two 'watering holes' restaurant bars that were our favorites. At the El Bogedon, we tried our luck at mastering the 'bota', a kidney-shaped Spanish wine flask made of goatskin. In order to drink the wine, the idea was to tip the flask up and away from the face so that the wine poured in a steady stream directly into the mouth without spilling it on your clothing. Quite a trick, and one that Glen was pretty good at, but at which I failed miserably, spoiling more than one outfit in the attempt. The waiters would encourage new customers to try it, and then stand by patiently waiting for them to fail and embarrass themselves, then laugh behind their aprons when they did. We gullible Americans were always fair game. The other café was the El Ranchito, which served excellent Spanish food and featured a really good live band that played only Spanish music. I knew the band members pretty well because I had often been there with Embassy friends. They liked to joke with me by stopping whatever they were playing to strike up my favorite song, 'Angelitos Negros' whenever I walked through the door. My fifteen minutes of fame, I guess.

Over our dating career I grew to know Glen's Squadron friends, most importantly his best friend Bob Galloway. Bob, along with Gary Grau, were officers whose wives were not with the unit, although Gary's wife Nancy did arrive at Sangley before the Squadron's move to Japan. Unbelievably, Nancy and Glen already knew one another because they had attended high school together in Iowa. Small world to be sure. Our dates on the Base were spent at the pool, a popular meeting place for us unattached types. Once Glen was out of circulation, the principal 'eligible bachelor' was Karle Naggs, whose happy personality was a magnet for the uncommitted young ladies. Because I lived in Manila, there weren't many opportunities for me to become well acquainted with the wives in the Squadron, but I do remember everyone was always friendly and welcoming. I admit, despite their warm acceptance, I felt a bit out of my element since they had established relationships — some of them long-term — and I was an unknown quantity. One evening, however, at some sort of social thing at the Base, Joan Angier drew me aside and said something like: It's nice; Glen is so much more talkative since you two have been going together. Although I did not consider him a 'quiet type,' he must have had the reputation for having little to say around the wives. Anyway, that may have been an offhand remark to Joan, but it gave me a nice little feeling of acceptance and I remember it to this day with fondness.

From the onset, I was more than just a little dazzled by the idea of dating a bona fide naval aviator, and I was naturally eager to hear all about the airplanes Glen flew as well as everything else about his work. To satisfy some of my curiosity, I was once permitted to watch from the BOQ area while he and a crew practiced 'touch and goes' a term I learned pilots used for practicing takeoffs and landings. It might have been a tedious though necessary exercise for them, but the fact that Glen could maneuver that big blue plane certainly impressed me. The closest I ever came to knowing anything about how the Navy operated on the Base was on that visit. Over the hours we spent together, I did notice that for some unspoken reason Glen did not share specifics about his Squadron's missions or any of his other duties. That was a bit puzzling. Nothing that was said or unsaid ever gave me the impression that there was anything mysterious or clandestine about what went on at Sangley Point.

The only time Glen ever shared any particulars about a mission was on an evening I will never forget. He arrived at the house after one of his "I might not see you for a couple of days" periods to which I had become accustomed. He seemed unusually solemn and I could see something was bothering him, because instead of casually visiting with my family, he asked if he and I could sit in the den and have some privacy. What was this going to be about? He hurriedly sat down on the couch in the den, took a quick breath and began to speak. What came tumbling out was an unbelievable, incredible story! I learned very quickly that evening that Navy wives (and prospective ones) must be prepared for literally anything!

I call his story "The Incident". It is impossible to relate it without explaining something of how airplanes work, how quickly events in those airplanes can put their crews in real danger, and how quickly well-trained individuals can handle a serious, even life-threatening situation. Much of the following description comes from Nancy Grau's husband, Gary, who was Officer of the Day at VQ-1 Base that night. I have also included some of the required personal report made to his superiors after the incident.

On the night of 27 July, 1955, Jim Edixon was the pilot and Glen the copilot of P4M-1Q 124362 on a return flight near the island of Okinawa-- when the aircraft's port engine abruptly DROPPED OFF into the Pacific! I didn't know much about airplane flight, but I assumed the loss of an entire engine would be disastrous and cause major panic in the cockpit.

> *Gary: "The P4M had two sets of engines: two propeller-driven that supplied the power for flight and the electrical power necessary for other*

systems, and two jets that were only used for auxiliary thrust if needed and was always on when landing, if needed. Prior to the episode the flight was proceeding normally. Then, suddenly the plane started to shudder and began a tight downward spiral to the left that pilots term a "spin in". When the crew noticed the engine on Jim's side seemed to have stopped, they followed standard procedure and tried to feather the prop, not realizing that it was actually gone. The objective then was to start both jet engines, however, feathering the prop unfortunately caused oil to spill into the left jet's air intake, meaning the jet sputtered repeatedly and would not start. The Plane Captain, a mechanic stationed on the plane's flight deck and leader of the enlisted flight crew, rigged a spotlight, shone it out the window behind the radar operator and reported, "The engine is gone!"

Immediately they stopped the feathering process. They were able to reduce the power to the right prop engine while starting the jet engine on that side. This probably meant that it was difficult to maintain direction, but they had reduced the severity of the spin. Until the right engine was started, there was not yet enough power to maintain direction or altitude, but they may have been able to slow down the rate of descent. It also allowed them to generate enough electrical power with the right prop engine, even though they had to reduce the power to it. Meanwhile, when they stopped the feathering procedure they had stopped pumping oil into the left jet engine. Once the feathering procedure ceased, the slipstream of air caused the freewheeling turbine in that engine to spin and eliminate most of the oil. The gas being pumped into the left jet engine was able to ignite the mixture of gas and remaining oil. Now they had the necessary power with the three engines to increase their air speed and be able to maintain both direction and altitude with judicious use of the throttles and flight controls. That allowed them to have electrical power which was generated with the prop engine on the right and enough control to make an approach to land."

Glen: "It has been my previous experience that GCI (Ground Control Intercept) stations often run practice intercepts on aircraft when entering Okinawa. For this reason, my first thought when the engine fell was that an interceptor had collided with us. An air to air collision is the most nearly correct description I could offer. The aircraft shuddered violently with an accompanying series of resounding noises. The weather at this time was intermittent IFR (Instrument Flight Regulation) and extremely dark due to the absence of any moon.

After the first indication, the airspeed dropped off rapidly and the plane ceased to respond normally to the controls. We entered a flat turn to the left at which time the ball on the indicator was deflected to the extreme.

About this time the aircraft started to descend rapidly (about 3000 ft/min) with an airspeed of approximately 120-130 knots in an extreme nose down attitude. The pilot told me to add full power on the starboard engine which I immediately did, however this did not appear to improve our rate of descent. It became obvious at this time that to alleviate the situation we would need power on our port side so I directed all my attention to a port jet start attempt. The pilot had already unsuccessfully tried a port jet start. I used normal air start procedures and although the port jet did not react normally (probably due to it being loaded with oil which was streaming from the port fire wall), it finally lit off. The jet start was effected at about 2000 ft. and we finally leveled off at approximately 1700 ft. During the entire time we were buffeting severely. I felt certain that we had lost a major portion of our aircraft but was thinking more in terms of part of the control surfaces (wing or tail) rather that an engine.

When the aircraft first started its rapid descent I felt certain that we would be forced to ditch, at which time I came up with a "mayday" report, followed by a prepare to ditch report to the crew. We proceeded to Naha (Okinawa) with continued buffeting. Landing was smooth and normal except for an extreme nose down attitude.

Acft (aircraft) settled on tail skid when we stopped. In summary it is my opinion the aircraft could not have maintained flight without the port jet operating. Moreover it appeared that a successful ditching could not have been effected because of the erractic response to the controls combined with slow airspeed and rapid descent.

Also, it was impossible to see the water due to darkness and clouds. Our landing lights were inoperative.

I include the "incident" in my story for two important reasons. First, I believe the crew of P4M-1Q 124362 were not atypical of the men of VQ-1. And this was not the only "incident" to occur during the Squadron's tour in the Far East. But it was theirs, and they worked together and demonstrated that well-trained personnel instinctively have confidence in their ability to handle a serious emergency with cool heads and live to tell of it.

The second reason is very personal. Glen said that, to his surprise, the idea that a person's life comes into focus when that life is in jeopardy was all to true for him. He said the overwhelming feeling he had as the plane headed for the ocean was of profound sadness because, "*It looks like Marilyn and I are never going to get married.*" I couln't think of anything profound to say. He was there, safe, and I was thankful. Later, upon hearing the news of the VQ-1 plane that was shot down with all hands lost, I tearfully thought about those wives and realized how fortunate I was not to have been forced to live out their story.

One evening Glen announced that he would be part of a crew that was to fly one of the planes stateside for its regularly scheduled maintenance and overhaul. We were serious about our relationship by then, and he asked me 'to please stay put' until he returned in two weeks or so. Those two weeks stretched out to be several, leaving me in a quandary about my lengthening visit, but definitely not wanting to return to the U.S. and jeopardize any chance of a future for us. Before he left he gave me a lovely little lacquered Japanese music box 'to remind me of him' while he was gone. It had the prescribed pin-skirted ballerina and played *'Shina No Yoru'* (China Nights). The current number one song heard everywhere in Japan in those days. For weeks, the household had a steady diet of that tune. (in researching the internet for the correct spelling and meaning of 'Shina' for this story, I found that many servicemen also considered 'She Ain't Got No YoYo' as many jokingly translated it back then — a lasting remembrance of their experience in the Far East)' I still have that music box, and though it has suffered somewhat over the years, it can still play a few chords of that cherished song.

In all good conscience I cannot say, "I moped around the house all the time Glen was away." Again, to help keep me entertained in his absence he introduced me to a group of English and Scots people he knew who were passionately devoted to warm 'spirits' and off-hours fun and games. Because they liked Glen, they graciously took me into their fold and included me in their good times. Saturdays, no matter the temperature and humidity we religiously played field hockey after which we drank warm beer. Monday evenings were spent at their private 'club' (the Brits always seem to have one of those regardless of where they are) where we danced Scottish reels and drank Bailey's Irish Crème, no ice. Other evenings we listened to music at one of their apartments and drank gimlets, no ice, or drove north of Manila to hear a particularly gifted Basque guitarist they had discovered. Once I was even enlisted by one couple to help them compete as part of their crew in a sculling race, but because I kept dangerously rocking the boat while shipping the oars they decided I was a woefully 'unworthy seaman' and gave up the idea. And so the days passed while strains of *'Shina'* echoed through the house.

Finally one evening a beaming Florencio met me at the door saying 'Gleen' had called and wanted me to meet him the next evening at the Army Navy Club in Manila. He was back! Wow! I don't know who was more excited, Florencio or I. I should add here that Florencio had a particular fondness for Glen as a person as well as a suitor and proudly considered himself our own matchmaker. At my prompting, he even taught me the phrase, 'inibig kita' which meant 'I love you' in Tagalog. The next day dragged on endlessly while I fretted over what to wear, changing outfits several times. It was first-date all over again. Walking down the long veranda that evening, I admit I was nervous. Had the extended separation changed our feelings for one another? And suddenly there he was. As I neared him, I realized something about him was definitely different. He had grown a MUSTACHE! In all my fantasizing about how the reunion would happen, I never ever imagined him like that!

All the while over dinner and the inevitable,

"What did you do all the time we were apart?" conversation, I was fearful of how I was going to break the news that I did not like mustaches, particularly on the one person I thought might be the love of my life? (It was sometime after we were married that he confessed the mustache was his last symbolic statement of freedom before ending his bachelor life forever, plus I think he secretly expected me to like it.) Anyway, we managed to get past that hurdle and proceeded to the serious business of reclaiming our relationship. It was a short couple of weeks later, sitting in the little black Ford in front of our house that he presented me with a beautiful solitaire, and I became engaged to my bona fide naval aviator.

Once we decided to marry in the Philippines, we were faced with a host of knotty problems, not the least of which was the fact that Glen's parents did not know me at all except for some snapshots and information supplied by him, plus a few friendly introductory letters from me. Frankly, I might be considered his proverbial 'war bride' the unmet person he would deposit on their doorstep with, 'Look what I found in the Philippines'. However, the situation understandably had to cause them some sleepless nights. I believe the confidence they had in their son and his sensible Midwest upbringing comforted them that he would not bring home someone totally unacceptable. Most assuredly, they were no doubt relieved to learn that the Navy wisely had its own protective parameters concerning servicemen who wanted to be married while on an overseas tour of duty. We were required to meet a few times with the Chaplain who would determine the seriousness of our intentions (actually mine, I believe) before giving us his blessing.

Some other information about our upcoming nuptials: I had no preconceived ideas about my 'perfect dream wedding.' Moreover, I was naïve and totally out of my comfort zone. Yet I was suddenly faced with making all those monumental, forever after decisions without the luxury of a bridal shop or a cake bakery down the block. What was I going to do about a Matron of Honor, announcements and invitations, trousseau, reception, flowers, a cake, and THE DRESS! Thankfully, it all miraculously came together. Jinny Martin, who was the closest Navy wife/friend I had, kindly agreed to be my Matron of Honor. That was a major relief since she was a calming influence and an enormous help, even to the point of designing her own dress and having it made without much input from me. To fill out the wedding party, Bob would be Best Man with Don Sumner and Karle as ushers.

THE DRESS was the result of much time spent poring over bridal magazines followed by multiple trips to the dressmaker and the fabric store. Because there was no such thing as a dress shop, Manila had an abundance of excellent seamstresses who create whatever the customer had in mind as long as he or she had the concept, the patience, and the pesos. Luckily, I already knew a dressmaker I felt confident could perform just such a miracle. However, true to my innate inability to make a decision, I took a long time choosing from the bolts and bolts of luxurious imported fabrics. When I couldn't procrastinate any longer, I finally settled on rich white brocade shot through with a silver thread that seemed perfect for the simple design I preferred. With the momentous dress decision finally made, I could breathe easier and move to the item on the checklist.

The last item on the agenda was the reception that would be held at the Officers' Club on the Base. They would also help with the food and the cake. Oh, the cake! At a meeting about the cake, the baker asked what topper I would like. Keep in mind I had few wedding prejudices but I did know I didn't want the usual bride and groom dolls, so he suggested maybe a pair of white doves might be nice. Uncreative to the core, I had no other suggestion and doves signaled peace and love, didn't they?

"Fine" I said, "doves it is." I left confidently checking off another decision. Early on, the wedding day I ducked in to see the cake masterpiece only to find the baker feverishly trying to touch up the white icing he had piped onto a pair of very large one-dimensional cardboard doves that could have been made by a local kindergartner. Gracious, did I agree to that? Well, it was too late to be disapproving or to settle for the bride and groom dolls. So, I had a good laugh and accepted what was the only flaw that day. My naivete had come home to roost (no pun intended). Fortunately, the cake tasted great, but we eliminated the cake-cutting pictures from our wedding photo album for obvious reasons.

Since a military wedding was an unusual occurrence on the Base, someone (I don't recall who) thought a military 'Arch of Swords' as we exited the chapel would add a classic element of tradition to the ceremony. An intriguing idea, but no one had the faintest notion of the correct procedure for such an event, and so there was a scramble to investigate what protocol had to be followed. Tradition stated that members must wear full dress whites, white gloves, be sure the swords were held sharp side up (no doubt to avoid accidentally slicing the couple) and one individual should give the command, "Swords Up," when the couple entered the arch. In addition, after the ceremony a designated officer should give the bride a modest 'sword swat' on the behind saying, 'Welcome to the Navy Ma'am.' Gary Grau, Fred Groenert, Fred Smith, Bob Hansen, Don Angier, and Wells Rathbun offered to be the Honor Guard. POOR Wells took a good amount of teasing for spoiling the dignity of the occasion because he could not raise his arm properly during practice due to a tetanus shot he had received. Our home movies of those practices show a determined group intent on performing perfectly, at the same time having a great deal of fun holding their own version of a 'wetting down ceremony'.

Shortly before the wedding, my mother and I heard a plane carrying U.S. supplies was coming into Clark Field, the Air Force Base north of Manila. We decided to make the long trip to the Base because it was rumored there would perhaps be some personal 'trousseau-like items on board'. Instead of driving my parent's mercury, we hired a car and driver because it was deemed unsafe for two American women alone to drive that distance through an area in which the Communist-led Hukbalahap rebels were active. Our driver accepted the job but had a set of specific rules in case we were stopped by anyone suspicious: Don't say anything, let him do all the talking, and have cartons of cigarettes handy in case a bribe was necessary. Predictably, we were flagged down twice by men in military fatigues who didn't look particularly menacing at first glance, but who sported some serious-looking weapons. The men waved us to the side of the road. Wearing large dark sunglasses, they peered into the car and questioned us at length, but we kept quiet and let our driver do the talking. (Filipino men seemed to have a penchant for large dark glasses.) They wanted to know where we were going and what we were going to do when we got there. Each time it was heart stopping, but per our marching orders, we kept quiet and let the driver, with cigarettes in hand, do all the negotiating. Thankfully, in both instances the men turned out to be police constabulary who were holding checkpoints due to reports of harassment in the area by suspected Huks. Both times, we were waved off with an apology when they determined we were not conspirators but two harmless American women on a shopping spree. Despite the underlying tension, we did manage to enjoy an otherwise leisurely drive through typically lush Philippine countryside even chancing a stop to sample some wonderful native fruit from a plantation roadside stand. In the end we navigated both legs of the trip safely and I was able to come home with two purchases; some perfume I particularly liked and a multi-layered flouncy purple taffeta slip that was irresistible because it looked 'trousseau,' but was something for which I had absolutely no use! Relaxing at home that evening we agreed it was an unforgettable experience with an interesting lesson in Philippine-style international diplomacy.

A word here about the wedding and shower gifts we received. People were unbelievably kind and generous to us, two people they knew very little or not at all in some cases. Some of the wives of Jack's associates at the hospital entertained me with a shower "tienda," or afternoon tea, a cultural tradition adopted from the large Spanish population that settled in the Philippines. It was a very dressy affair with iced spiced tea, miniature lumpia; dainty sandwiches and small sweet cakes served al fresco on the hostess's veranda. Hands down, the most popular gift I received that day was a set of placemats. There were no less than fourteen of them, each one a different color with a different native motif or design. Some of our wedding gifts, I realize now are irreplaceable not only because of their sentimental value. For example, we received an amazing carved Philippine mahogany wood table depicting a scene from peasant everyday life and a beautiful Nara wood punchbowl set. Both of those woods are highly prized because they are so unique, the most valued Philippine mahogany is raised on Luzon, and Nara is so precious it is now restricted for sale outside the country. But the gift that to this day always makes me smile is one that — Jinny again — thought up, a rolling pin signed by all the wives and inscribed 'good rollin.' Because it was varnished to preserve the signatures, I used it sparingly and it served me well over the years until I decided to retire it as a keepsake.

One event we had not planned for was a typhoon that slammed the area the week before the BIG DAY. According to sources after the fact was termed a 'high typhoon,' which meant it was more of a tropical depression disturbance with more moderate winds than a much more destructive 'low typhoon'. Regardless of the terminology, I remember huddling in my bedroom for hours that night listening to the howling outside the wooden window slats on Carlos' well-designed house would stand the test of the driving rain and the 125 mph winds. Above all, I was fearful our entire wedding might become a casualty of any resulting chaos. To my relief, city damage was minimal and our backyard cleanup the next day only involved cleaning away palm fronds and other debris that had blown in from the surrounding area. A call from Glen said all was well at Sangley. The wedding was still on track.

Except for one small hitch on Thursday, I was home alone when, of all things, two constabulary officers appeared at the front gate strongly 'suggesting' I accompany them downtown to the police station because of some question concerning my stay in the country. Not understanding what possible crime I could have committed, I decided a belligerent argument over an immediate explanation was unwise and so I got into their car. I sat quaking in the back seat — in custody, I suppose — while they sat in the front seat laughing and jabbering in Tagalog. Was that about my fate? At the station, I learned I was on a detainee list for having an expired visa, a very serious breach of diplomatic policy I was sternly made to understand. Oops! In the wedding confusion, that one never crossed my mind. After I hastily explained that I would be married to an American naval officer in three days, thus changing my status, what followed was a flurry of phone calls to verify if my excuse was legitimate. I waited nervously balancing on the edge of my seat, while visions — not of sugarplums — but of lawyers, detention, even deportation, 'danced in my head!' Finally, after what seemed like an eternity, an officer assured me that the navy had vouched for me and I would be released under a temporary visa. Then there was a furious, stamp, stamp, stamp, stamp of one official document after another and I was politely escorted home. It's an amusing story in the telling, but certainly not amusing at the time.

My wedding day finally arrived and it was an exciting blur of images that I thought at the time would be indelible, but which have unfortunately faded or become unclear over the years. I know Don and Alice Sumner kindly offered their house as a place for my family and me to headquarter before the zero hour of 3:30. (I don't know why I chose such an odd hour to get married; I suppose it was my attempt at being worldly, but in retrospect, it just seems quirky and pretentious). It was a long afternoon. Then, at 3:30, all the minor frustrations about cakes, dress patterns, and typhoons faded away as Jack and I stood looking down the chapel aisle toward the front where Glen was waiting. I was about to become a Navy Wife.

Probably the most noticeable result of the typhoon was the excessively sultry air it left in its wake for the next few days. Since our wedding chapel was a Quonset hut without benefit of air-conditioning, the ceremony was not without some discomfort for everyone attending. I was in my simple cap-sleeved gown, but Glen wore sharkskin dress whites, with a high collar that fit snugly around his neck, hardly the most comfortable outfit in that heat. Laughingly, my mother always said Glen's face was so flushed during the ceremony that she fully expected to see steam rise from a vent in the top of his head. (I always contented myself with the belief that it was the heat and not second thoughts about the "I do's) The ceremony was the short version — I recall smiling nervously through most of the chaplain's words — and before I had time to catch my breath, we were pronounced Lieutenant and Mrs. Glen Carlstrom and were walking down the short aisle to the back of the chapel. When we reached the outer door, the guys snapped to attention, the swords went up (even Wells), and we passed under the arch. It was awesome! It was unforgettable! After some backslapping for a job well done and some joking around about who would have the dubious honor, I think it was Don Angier who finally gave me the 'swat'.

The outside patio reception was lovely. Guests included acquaintances from the Embassy group, hospital associates of Jack's and their wives, the English and Scots, and our VQ friends. The whole affair had a uniquely international flavor, with the Filipino men in their dress Barong Tagalogs and some of their wives in their colorful high-sleeved Baro at Saya dresses. Champagne flowed and we cut the CAKE as a finale. Following the reception, we ate a late wedding supper at one of Manila's most popular restaurants, enjoying cheese soufflé and Amontillado sherry — an anniversary menu tradition we continued for many years. Our wedding night reservation were at the Bayview Hotel in Manila, since we faced a long drive the next day to the place where we would spend the next week. Upon checking into our much-anticipated "honeymoon suite," we were shocked to find they had given us a room with twin beds! For a while, we stood looking at each other, not knowing quite how to handle the awkward situation while the confused bellboy stood on one foot and another, patiently waiting for his tip. Finally, after we agreed the arrangement was totally unacceptable, the bellboy, again trailing us with our luggage, followed us back down to the lobby where Glen spent several intense moments insisting as diplomatically; as he could that a room change was in order. He won, we moved!

We spent our honeymoon in Baguio at Camp John Hay, the military R&R mountain resort. Our 'cottage' has a two-story affair off by itself that had a kitchen and could accommodate quite a number of people if need be because the upstairs sleeping room had no less than five double beds! Goldilocks would have had a nervous breakdown! It was definitely not the Ritz Carlton, but it also had a cozy fireplace that served us well during the chilly mountain evenings. Days were spent doing touristy things such as a trip to the bustling Baguio open market. I was intrigued by the vendors, who surprisingly displayed their native woven yarn goods, not the usual tables but on long poles suspended from the rooftops. Those yards and yards of color swinging in the breeze were irresistible to me. I had to have some. I still have that material just as I bought it that day. I never had the heart to spoil it by cutting it up. Moreover, native woven yardage was much narrower than the 36 inch widths American were accustomed to, so making anything wearable wasn't an easy proposition. The other popular tourist attractions were the stalls where the local woodworkers carved native ironwood and Nara into an endless supply of carabao statues, wood trays, and horse head bookends. Every stall had the same expertly crafted selection, so it was difficult to make a choice. After considering all the possibilities, we finally decided a set of ironwood Igorot heads was the perfect souvenir since that tribe was native to the Baguio region. The rest of the time, we just enjoyed the spectacular scenery and the thankfully cooler mountain air.

When we returned from our honeymoon, Jinny Martin came to the rescue once again by offering their spare bedroom to us for the week before we all left for the squadron's move to Japan. The Martin's house was outside the base in Cavite City, and Jinny's expertise and passion for collecting native art had transformed a less than perfect house into an entirely livable home. While there, to show my appreciation for their kindness, I suggested I prepare Sunday brunch. I can't remember what else was on my menu, but cheese soufflé — hopefully a recreation of the dish we had for our wedding supper — was the main attraction. Armed with a cookbook and everything I could ask for in the kitchen, I felt confident I could bring the whole thing off with no problem. So, I sent Glen and the Martins to laze about assuring them my meal would be ready in jig time. As the minutes dragged on, I tried to smile confidently while I checked and rechecked the oven countless times. I learned the hard way that cheese soufflé cooks on its own timetable. It cannot be rushed. Two and a half agonizing hours later we sat down to eat! It was a humbling initiation for me as a novice cook.

Just before we left Sangley for Japan, I stopped by the house to check in with my Mom. Assuming my new role as an organized WIFE person, I announced that the Navy had successfully rounded up all our wedding gifts and packed them for shipment to whatever base would be our new home in the States after we left Japan in December.

"Isn't that nice," Mom said with a wry smile, "but I'd like to see how the Navy is going to pack the present that arrived this morning!" I looked to where she was pointing out in the breezeway, and there stood a bed! A real Filipino double bed! It had a rattan headboard and a flat frame made of rattan that served as a box spring/mattress. That was it. There was no fluffy quilted mattress such as one would expect, so stretching out on it was like relaxing on a concrete slab. It was a generous gift, and the giver no doubt thought a bed was the ideal present for a pair of newlyweds, but what were we going to do with it? More importantly, what was the Navy going to do with it at this late date? However, true to Navy efficiency, they picked it up, added it to our shipment, and we used it for years as a humorous anecdote, but never as a bed.

Moving to new quarters in Japan was exciting, but sadly, that meant saying goodbye to my family who I would not see for an uncertain length of time. Work on the hospital was still far from completion. Aside from that, the plane trip (although I don't remember how we flew) had me again in a whirl because changes in my life kept occurring so rapidly that it was hard for me to take it all in. I just kept telling myself that at last I was going to have my new husband all to myself in our first real home. As it turned out there was not enough housing for all the families from Sangley to live on base in Iwakuni several of us were sent to Nijimura. It was an Army housing complex near the town of Hiro, a rice-growing district. Our quarters were two-story units of four apartments in each building. Our apartment was on one end of the building and the Grau's were on the opposite end. Decorated in 'basic military' it was up to each family to use whatever personal belongings they had to make their quarters all their own. Glen and I had none of our wedding gifts to place about and we had a frustratingly short time remaining before his rotation, so it seemed useless to spend much time decorating.

Marilyn and Japanese Christmas Tree, 1955.

On the first morning in what could be called our very first home as newlyweds, I was awakened by ominous noises coming from downstairs. Looking down from the top of the stairs, I was surprised and relieved to see that our intruder was a pleasant looking Japanese lady bowing and smiling up at me. "Okusan," she kept calling me, still smiling and bowing. I finally surmised that 'okusan' must be a Japanese term of respect for the 'lady of the house' and she was the housekeeper we were inheriting from the previous tenants. Wait a minute! I was a new bride who wanted to have my house to myself and prove what a great homemaker I could be. I certainly did not need a housekeeper. Regrettably, I cannot recall her name, but I found her to be pleasantly cooperative, so I decided to keep her on, in part because that was probably her only reliable source of steady income. Since Glen was gone the whole day, leaving me alone in the house, the main difficulty I had was finding enough work to keep her busy. She washed the few dishes that I purposely left from our previous night's dinner. She dusted what didn't need dusting, and she washed my personal laundry by hand each morning, one of my more bizarre luxuries. I consequently did nothing except make an effort to stay out of the way, and it was a daily challenge to find some unique way to fill my time. One day I even resorted to lining up all of our shoes and armed with rags and polish proceeded to give each pair a first-rate shine.

Our gadgets and we Americans must have been a constant puzzlement to the Japanese. Like the plan-ahead person he was, Glen came home one evening announcing that he had bought a new Philco refrigerator at the Iwakuni exchange in anticipation of our imminent move back to the States. Due to a mix-up in the shipping arrangements, it was not left on the dock for U.S. delivery but was instead dropped off at our apartment by mistake. Not being able to stand the suspense, I asked the deliverymen to open the crate. And there it was my very first appliance, a beautiful gleaming white refrigerator complete with the latest example of ingenious American design. A gold V-shaped door handle that allowed the door to be opened from either direction! Standing majestically on the sidewalk all that afternoon, no one could resist a curiosity. Everyone who gathered around, Japanese and Americans alike had to try their hand at opening and closing the marvelous wonder door. You might say it became an unlikely ambassador at-large of friendly Japanese/American relations that day. In the end, it was again crated, and eventually found its way into our various stateside homes where it lived for several years as a whimsical reminder of that afternoon in that far off land.

Our stay in Japan was filled with many 'firsts' not the least of which was my first Thanksgiving dinner. I cannot imagine why I opted for the challenge of a complicated dish like duck l'orange as the perfect way to start our holiday tradition, but I did. My inexperience as a cook surfaced again. Add to the menu petite poi's (fresh peas to me) in lettuce leaves that would be prepared in the shiny new copper chafing dish I purchased from a Japanese vendor at the Officers' Club. All morning I laboriously peeled and sectioned oranges, readied my peas, and finally slipped my carefully trussed duck into the oven. After all was set, Glen and I then spent all afternoon visiting other families in the apartment complex who had graciously invited us to share in their holiday celebrations. Finally, around seven we returned home and sat down to eat my holiday dinner. Two unfortunate things happened: first, my seriously overcooked duck, resting on an unexpected inch of grease, literally collapsed under the first touch of the carving knife, and second, upon firing up my chafing dish, a cloud of black sooty residue began floating through the air because I had used alcohol for fuel instead of the prescribed Sterno. Believe me; our patient housekeeper had more than enough dishwashing and cleaning to keep her busy the next day!

On occasions, we wives accompanied our husbands on the boat to Iwakuni for a day of shopping at the Exchange or for various appointments. The trip was an exciting opportunity to get a glimpse of yet another expect of Japanese life apart from the houses and truck gardens that lined the road outside the Nijimura Base. From the boat, we could see the distant shoreline was dotted with wooden frame houses and moored fishing boats that were typical of a hardworking rural class. The boats we saw out to sea, we learned years later, were 'fishing' all night, but not necessarily for fish for the dinner later. Because it was late autumn when we were there, the countryside had a drab color palette of grays and browns, although beautiful in its own right, there were no delicate flowering cherry trees so typical in Japanese art forms. In the same way Japanese interiors and everyday dress mirrored those same earth tones, a reflection I would guess of their inherent preference for serenity and simplicity in all things. Absent were the electric reds and yellows so representative of goods stamped 'Made in Japan' that eventually flooded the post-war market. I did see some of those goods firsthand in a Hiro shop that sold novelty toys of all descriptions. In contrast to all that gray and brown, it was a riot of colors. In the front of the store, no doubt prominently displayed because of its sales potential, was a windup monkey in a red hat that lewdly bucked up and down loudly clacked cymbals. I'm not a prude, but for some reason I thought it was disgusting. However, it was certainly a popular toy with everyone else and I saw it everywhere. There is no understanding public preference, but fifty-six years later that silly monkey is still on the market.

Traditional costumes were anything but drab. They were exquisite colorful worn by true geishas or on the stage in Noh performances. Hiro had a geisha house where several couples from our complex decided to experience an evening of Japanese style dining. I'm not certain whether the women at the house were geishas in the truest sense or hostess/entertainers who followed traditional customs and wore traditional geisha dress. It was, of course, a 'sit on the floor dinner' with ethnic food, sake, and after-dinner fun and games. During the after dinner game portion of the evening I remember we played 'pass the orange' which meant passing an orange from under the chin of one person to another without dropping it on the floor. What followed was a lot of fumbling (by us) and giggling (mostly by the geishas, and it felt somewhat like being in the first grade over again. One had to forget feeling silly and surrender to the spirit of the evening. Mind you, our hostesses were gracious and saw to our every need. Overall it was a pleasant evening except for one not-so-minor detail. In the geisha way, they were overtly more attentive to the men, with shyly provocative glances and even the exclusive 'men only' offer of a backrub or two. To be sure, the men smiled and happily accepted the attention, much to the annoyance of us wives who had to grit our teeth and accept it in good grace.

The most important 'pilgrimage' Glen and I wanted to make before we left Japan was a side trip to Hiroshima to see the results of the atomic blast. As Americans we both felt a deep sense of responsibility to see for ourselves what had taken place there. Consequently, one Saturday morning I took the boat to Iwakuni and we then boarded a typically overcrowded train for the trip halfway around the Inland Sea. Over the years I have told the story of that experience many times, always commenting that we felt like 'the only Caucasians in town that day.' I think we were. It was impossible not to feel conspicuously alien and ill at ease, particularly once Glen was in the uniform of the country responsible for the destruction. Naturally we wanted to see the center of the blast area, but no one we approached would (or could) give us directions or make an effort to understand us, so we gave up trying and just wandered around the area. What we saw was a kaleidoscope of strange, sometimes shocking images. Ten years after the blast the streets were still heaped with rubble as though the event had happened only recently. Most of the people on the street wore a white mask that covered the nose and mouth, probably as a protection against the persistent dust from the loose debris. A 'honey-bucket wagon' (the term for an open wagon that collected human waste for fertilizer) was being pulled, not by a truck but by horses that could more easily manage the rubble and the narrow streets. We passed by a commercial building with a small wrought iron fence, part of a memorial erected around the image of a person's outline that was literally etched into the stone as a result of the intense heat from the blast. It was horrifying. We just stood there silently staring at it while people passed behind us. I often wonder what those people felt as they watched us standing there. The sole exception to the overall feel of devastation was a funny little man in a red suit and jingling hat that seemed absurdly out of place as he marched up the street pounding on a bass drum and clanging cymbals to attract attention to something he was advertising. It was a curious testimony to the fact that life goes on even in the face of disaster. Waiting in the station for the train to Hiro later that afternoon, we didn't seem to want to talk about what we had seen – it was just such an overwhelmingly sobering experience.

The day finally came for us to leave. For me, it was difficult to say goodbye to Nancy and all the friends who had accepted me as one of them and who were a part of the most important event in our lives. For Glen it meant leaving friends with whom he had served for two years in more often under what we wives discovered much later, were the most stressful of circumstances. Everyone who left the Squadron received a beautiful wood tray as a parting gift. Ours was inscribed, 'Glen and Marilyn Carlstrom Mar. 53 – Dec. 55, VW-1A, VW-3A, VQ-1 Sangley Point, Iwakuni' surrounding a depiction of the Kentai Bridge at Iwakuni. It is a treasured remembrance that has had a prominent place in our home for all these years.

The first leg of our journey home began in Tokyo where we spent a day or two before our plane left for Hawaii. As I recall, we stayed one night in the Imperial Hotel, just to say we did, although not in one of their more luxurious rooms. Ours was small and the window looked out over the rooftops. As a diversion we took a shopping trip to a Japanese department store. In stark contrast with our experience in Hiroshima we did not feel in the least alien or uncomfortable. On the contrary, the salespeople were friendly and helpful and there was even a small group of happy, curious young school age children who followed us around from counter to counter, seemingly fascinated with everything we did. The next day we took a side trip to Kamakura to see the Diabutsu (the Great Buddha) and the Shinto shrine there. To see the Great Buddha is an understatement because that experience made such a lasting impression on me. It was amazing! Despite the Buddha's imposing size gazing at the face in repose, the quiet position of the hands, and the slight forward tilt of the body gave me a profound feeling of serenity that I have never forgotten. To this day I am forever thankful that my last memory of being in Japan was not of a desire for retribution, but of something powerful and positive.

In Hawaii, we spent a week at Fort DeRussy, the military R&R camp on Waikiki, little knowing that we would one day spend two years north of there at NAS Barber's Point and that our first daughter would be born at Tripler Hospital. As it turned out, our next assignment after Iwakuni was NAS San Diego where we would happily be reunited with Bob Galloway, and Don and Joan Anger. As a special surprise, I would learn to know Bob's wife, Nan, who I will forever hold in the highest esteem because, aside from being a great person, she actually ironed her bed sheets! The Galloway's and we saw a good deal of one another and spent many happy afternoons mixing concoctions in our wedding present blender. Our rented beach house was across a slough from a naval aviation facility that used drones when training pilots and flights usually passed right over our house. Sadly, one day I was watching when a plane and its drone accidently touched wings and both dropped into the ocean a short distance offshore. Authorities searched all afternoon, finally recovering the pilot who didn't survive because he could not get out of the cockpit. Even though I did not know that pilot, it was a loss all the same with other loved one left behind to grieve.

On our relocation to Hawaii, Glen's assignment was not as a pilot, but instead was as an instructor at a survival training camp. One of the horrible realities of the Korean conflict was the enemy's practice of brainwashing prisoners. The survival school was the military's effort to prepare American servicemen to cope with possible capture and indoctrination. Our two years in Hawaii were the last of our adventures overseas. Corpus Christi, Texas, then became our last base before Glen retired from active duty and joined the Reserves.

How do you adequately describe events that define the rest of your life? For me, those pivotal events began with a flight across the Pacific and ended at the back door of a navy chapel in the Philippines. In a sense, I owe the Navy everything because it gave me Glen with whom I spent forty-six years — including the births of four children — until his death in 2001. During our navy years, we lived in many places and met a host of nice people, but the friends we made in VQ-1 without question have remained the most valued!

The goal of many organizations — the military in particular — is the establishment of an 'esprit de corps' a devotion to shared purpose and commitment that any successful group adopts in order to survive. The Navy wives I knew had that. No matter what the circumstance they supported one another and they supported their husbands in whatever way was necessary. The fact that they were unaware of just how important that support was at the time is relevant now. They survived. They were smart, capable, caring, fun loving, talented, inquisitive, trustworthy, tolerant, strong, adaptable, and wise. They were Navy wives in the best sense, and I am proud to have been one of them.

Marilyn Carlstrom

Autobiography

May 17, 2011

Washington DC is my hometown. Although my family would have been considered economically disadvantaged, my younger days were spent happily taking full advantage of the free historical richness that was Washington. Otherwise, I led a pretty mundane young adult life. I graduated from high school after a move to California, made two unsuccessful attempts at college and then went to work.

Everything changed for me in 1954 when I found myself in the Philippines visiting my parents and there met and married my naval aviator husband Glen. Two of our children were born during our military life: a daughter Karen in 1956 while we were stationed at NAS Barber's Point in Hawaii, and a daughter Sandra in 1958 while we were in Corpus Christi, Texas.

When our Navy life ended, we settled in Emmetsburg, Iowa, where Glen operated an oil distributorship and I spent my free time volunteering on projects and serving on committees. Our other two children, John born in 1960, and daughter Susan, born in 1962 are 'post military service' Iowans. In 1975, we also welcomed our Swedish 'son' Lars, who came to us as an exchange student and remains an integral part of our family. We have eleven grandchildren.

After Glen retired from business, the carefree life of travel and adventure we had envisioned was brought short when he lost his battle with cancer in 2001.

In 1981, at age 50, I decided to go back to school, earning an Associate of Arts degree at Iowa Lakes Community College in Emmetsburg. I am now employed as a proctor at the school, a position I genuinely enjoy since it affords me the opportunity to be in the classroom and establish relationships with students and instructors.

The Iowa Lakes mission statement espouses lifelong learning, a philosophy I enthusiastically support because my education experiences, past and present, have greatly enriched my life.

APPENDIX
Glossary of Terms and Abbreviations

A = ABLE

ABLE = The letter A; International alphabet in place in the earlier part of the enclosed stories. It later became ALPHA

AFB = Air Force Base

Air Force Base = AFB

Airborne Early Warning Squadron THREE (DETACHMENT ABLE) = (VW-3) Det Able

ALPHA = The letter A; "ALPHA" in modern Navy-ese. It replaced the earlier "ABLE" IN THE International alphabet in place in the later part of the enclosed stories.

AT1 = Aviation Electronics Technician First Class, for purposes of the enclosed stories. Elsewhere it would be Aviation Technician First Class

AT3 = Aviation Electronics Technician Third Class, for purposes of the enclosed stories. Elsewhere it would be Aviation Technician Third Class

Aviation Electronics Technician = AT

Bachelor Officers' Quarters = BOQ

BOQ = Bachelor Officers' Quarters. Housing for unmarried men or men unaccompanied by wives

BX = Naval Base Exchange (BX); also called Navy Exchange or Ship's Store; a sort of department store exclusively for naval personnel and designated others.

CAT = Civil Air Transport Airlines, founded by General Clair Chennault of WWII Flying Tiger fame.

Chief of Naval Operations = CNO

CNO = Chief of Naval Operations

CO = Commanding Officer

Commander Naval Airforce Pacific = COMNAVAIRPAC

COMMANDER NAVAL FORCES PHILIPPINES / COMMANDER IN CHIEF PACIFIC / REPRESENTATIVE PHILIPPINES = COMNAVPHIL/CINC-PACREPPHIL

Commanding Officer = CO

COMNAVAIRPAC = Commander Naval Airforce Pacific

COMNAVPHIL/CINCPACREPPHIL = COMMANDER NAVAL FORCES PHILIPPINES / COMMANDER IN CHIEF PACIFIC / REPRESENTATIVE PHILIPPINES)

Construction Unit Battalion 1 = CUB1

CUB1 = Construction Unit Battalion 1

Electronic Countermeasures Squadron One = VQ-1

Fleet Logistic Air Wing = FLOGWING)

FLOGWING = Fleet Logistic Air Wing

Geedunk = Navy snack bar

GCA = Ground Control Approach

GCI = Ground Control Intercept

Ground Control Approach = GCA

HMS = Her/His Majesty's Ship

ICS = Internal Communication System

IFR = Instrument Flight Regulation

LCDR = Lieutenant Commander

Lieutenant = LT

Lieutenant Commander = LCDR

LT = Lieutenant

LTJG = Lieutenant Junior Grade

Marine Corps Air Station - MCAS

MATS = Military Air Transport Service

MCAS = Marine Corps Air Station

Military Air Transport Service = MATS

Military Facilities nomenclature

The Army's may be a Post, Camp, or Fort. Their exchanges can be a BX or a PX, or just Exchange.

The Navy has Naval Bases or Naval Air Stations.

The Marine Corps has Camps or Air Stations
Military Payment Script = MPC
Military Sea Transportation System = MSTS
MPC = Military Payment Script
MSTS = Military Sea Transportation System
NAF = Naval Air Facility
NAS = Naval Air Station
Naval Air Facility = NAF
Naval Air Station = NAS
NMI = No Middle Initial
Okinawa ADIZ = Air Defense Identification Zone
Philippines; Philippine Islands = P.I.
P.I. = Philippines; Philippine Islands
Port Hueneme is (was) a Seabee (Civilian Battalion and Naval Base)
Post Exchange = PX. Army equivalent of Navy's BX
PX = Post Exchange. Army equivalent of Navy's BX
R & R = Rest and Recreation
SAR = Search and Rescue
Ship's Store = Naval Base Exchange (BX); also called Navy Exchange; a sort of department store exclusively for naval personnel and designated others. Although strictly speaking it should be on a ship, many people call the one on land by that name.
Snack bar (Navy & USMC) = Geedunk, derived from the sound made when pulling the lever on a coin-operated candy machine.
UHF = Ultra High Frequency
USNS = United States Naval Ship, owned by the Navy but operated by a civilian crew.
USS = United States Ship which is owned and operated by the Navy
VQ-1 = Electronic Countermeasures Squadron One
VW-3 Det Able = Airborne Early Warning Squadron THREE (DETACHMENT ABLE)

END OF AN ERA

In December 1970 an official Navy announcement was made regarding the closure of Sangley Point. The base changed from "active" to "inactive" status on July 1, 1971.

CAPT Waldo Atkins, USN, was the man responsible for getting Sangley Point ready to leave the Navy. He and seven officers commanded nearly a hundred sailors who would prepare the base for transfer. In a two-month period that spanned the rainy season and two typhoons, they stripped furnishings from 400 industrial buildings, moved stilt houses to the Subic Bay Naval Reservation and secured all government quarters and the remaining Sangley Point Buildings.

A total of 2,500 tons and 1,500,000 cubic feet of material assets were identified, packed and shipped by sea and land to various other U.S. military bases. Much of this transfer was accomplished at night and on weekends due to severely restricted barge and trucking schedules.

On September 1, 1971, the base was officially turned over to the government of the Republic of the Philippines ending 74 years as a naval facility. The U.S. Navy's presence provided substantial bookends around three years of occupation by the Imperial Japanese Navy.

It was the home to generations of Navy families like the ones you just met.

Sangley Point is currently used as a facility of the Armed Forces of the Philippines, although Nancy Grau has heard that the former flight line is now populated with luxury condominiums.

Back in the Navy, the Fleet Air Reconnaissance (VQ) mission over the last fifty-seven years has been accomplished by two proud squadrons: the VQ-1 *World Watchers* in the Pacific and by the *Rangers* of VQ-2 in the Atlantic and Mediterranean Seas.

Many pilots and crew did tours in both squadrons. In May of 2012, the OPNAV Staff announced that the Rangers would conclude more than 57 years of distinguished service in an official disestablishment ceremony at Naval Air Station Whidbey Island, WA.

The guest speaker was retired Capt. D.C. East, former VQ-2 Commanding Officer, and source of much of the information contained in the history of the squadron at the beginning of this book. Originally home-ported at NAS Port Lyautey, Morocco, VQ-2 executed a homeport change in January 1960 to Naval Station Rota, Spain. More recently, VQ-2 executed a second and final homeport change to NAS Whidbey Island in September 2005. The cost-saving move co-located VQ-2 with its sister squadron, VQ-1 World Watchers, increasing training and readiness.

While the *Rangers* have departed active service, the VQ-1 *World Watchers* continue the critical — and dangerous — Electronic Warfare, Intelligence, Surveillance, and Reconnaissance mission world wide in response to operational tasking by Joint Chiefs of Staff and Theater Commanders.

The families of the crews are also still on watch.

Sunset over Bataan from Sangley Point, 1955.

Down Memory Lane

Sangley Point Officer's Club

Ruins of Manila Cathedral

WW II damage to Cavite Cemetery Gate

Ruins of Santiago Prison

Japanese Santa

Made in United States
North Haven, CT
10 November 2026

22088257R00153